COMMUNICATION SATELLITES:
Power Politics in Space

The Artech House Telecom Library

Telecommunications: An Interdisciplinary Text, Leonard Lewin, ed.
Telecommunications in the U.S.: Trends and Policies, Leonard Lewin, ed.
The ITU in a Changing World by George A. Codding, Jr. and Anthony M. Rutkowski
Integrated Services Digital Networks by Anthony M. Rutkowski
The Deregulation of International Telecommunications by Ronald S. Eward
Machine Cryptography and Modern Cryptanalysis by Cipher A. Deavours and Louis Kruh
The Competition for Markets in International Telecommunications by Ronald S. Eward
Teleconferencing Technology and Applications by Christine H. Olgren and Lorne A. Parker
The Executive Guide to Video Teleconferencing by Ronald J. Bohm and Lee B. Templeton
World-Traded Services: The Challenge for the Eighties by Raymond J. Krommenacker
Microcomputer Tools for Communication Engineering by S.T. Li, J.C. Logan, J.W. Rockway, and D.W.S. Tam
World Atlas of Satellites, Donald M. Jansky, ed.
Communication Satellites in the Geostationary Orbit by Donald M. Jansky and Michael C. Jeruchim
Communication Satellites: Power Politics in Space by Larry Martinez
New Directions in Satellite Communications: Challenges for North and South by Heather E. Hudson
Television Programming Across National Boundaries: The EBU and OIRT Experience by Ernest Eugster
Evaluating Telecommunication Technology in Medicine by David Conrath, Earl Dunn, and Christopher Higgins
Measurements in Optical Fibers and Devices: Theory and Experiments by G. Cancellieri and U. Ravaioli
Fiber Optics Communications, Henry F. Taylor, ed.
Mathematical Methods of Information Transmission by Kurt Arbenz and Jean-Claude Martin
Telecommunications Systems by Pierre-Gerard Fontolliet
Techniques in Data Communications by Ralph Glasgal
The Public Manager's Telephone Book by Paul Daubitz
The Cable Television Technical Handbook by Bobby Harrell
Proceedings, Conference on Advanced Research in VLSI, 1982, MIT
Proceedings, Conference on Advanced Research in VLSI, 1984. MIT

COMMUNICATION SATELLITES:

Power Politics in Space

Larry Martinez

ARTECH HOUSE, INC.
610 Washington Street
Dedham, MA 02026

International Standard Book Number: 0-89006-167-X
Library of Congress Catalog Card Number: 85-047746

**This book is dedicated to my parents —
that crystal radio did it!**

CONTENTS

PREFACE

The runners charge on each other's heels as they turn the final corner and start down the tunnel leading into the Los Angeles Memorial Coliseum. The TV cameras mounted on trucks with their own microwave transmitters have followed the competitors in this Olympic marathon from the start, more than two hours earlier. Now the larger cameras located in the press booth focus on the lead runners in a dead heat for the finish line. There has never been a marathon finish like this in modern Olympic history, and never has it been watched like this one either. The runners make their break.

The crowd, already on its feet, watches the race on the two large screens at the far end of the Coliseum, seeing what entire nations of TV viewers thousands of miles away are watching simultaneously. A woman in Seattle punches the record button on her video cassette recorder to save the moment for her husband who is away at work on an oil platform, although he, too, is watching it by virtue of a satellite dish antenna mounted on the drilling rig, which is located hundreds of miles from the nearest telephone lines or television stations. Children in Portugal are shrill with excitement, shouting encouragement to their hero on the glowing screen as he vies for first place. In Africa, a young boy will soon run the thirteen kilometers back to his village with the news that his cousin is also a champion, news that has no other way of reaching the village except on foot.

For a single moment, the world is tied together, not so much in athletic competition, but in information and telecommunications services. A moment that shows not only the possibilities for the future, but also the discrepancies of current national technological levels. This is a book about these times, and the times to come with many of its fantastic and exciting changes, which we shall see before today's youngsters will even compete in the next Olympics. This is a book that examines the international political turmoil and conflict caused by one such information technology.

Researching and writing this book was also an adventure, both into outer space and to many parts of this planet. The research travel of more than 50,000 miles (equivalent to one round-trip to the geostationary orbit) was made possible with grants from the Deutscher Akademischer Austauschdienst and the Academic Senate of the University of California, Santa Barbara. Many thanks to the Institut fur Weltwirtschaft, Kiel, and to the Institut fur Luft- und Weltraumrecht, Cologne, for their assistance. *"Merci beaucoup"* to

Secretary-General Richard Butler and to the librarians at the ITU Archives in Geneva. *"Muchas Gracias"* to the Political Science Departments of the University of California, Santa Barbara, and California State University, Fullerton, for their support. I am especially indebted to Professor Carl Christol of the University of Southern California for his suggestions and advice. Appropriately, this book about information technologies was written on several computer word processors with assistance from Rose Mucci, Becky Davis, and Bill Hyder at UCSB, and Charles Sowers and Michael Marcinkevicz at CSUF. Jay Martinez was a thought instigator in the realm of information theory.

Nonetheless, the most rewarding and enjoyable part of this study was becoming acquainted with the people involved in the many exciting aspects of outer space exploration and telecommunications. Andrea, Jane, Juergen, Kim, Mahindra, Olga, Rich, Tony, and many others, transformed my earlier trepidations concerning field research. *"Vielen dank"* to the members of the Helmut and Hanne Hermann family in Stuttgart, Mainz, Hamburg, and Berlin for many happy times in Germany. Regardless of the language used, I have tried to portray the various perspectives in the debate over communications satellites as accurately as possible; any misrepresentations are clearly my own and not of the people mentioned above.

Finally, communications satellites have allowed people to look at their earth as it really is — a blue, white, and brown celestial home — a perspective of common destiny on a planet whose stewardship befits the higher qualities of any organism in this universe that looks out at a night sky and wonders in awe.

INTRODUCTION

> Call it the *Information Age*. Consider ours the *Information Society*. No longer a mere merging of computers and communications. No longer a coming together "one day." No longer the "Future." The future is now — the manifestation of "technological hop" that is producing fundamental innovations as fast as society can absorb them . . . We have within our grasp the most powerful productivity-enhancing force, the most powerful force for improving education, the most powerful force for lifting our lives. Embrace that force and we prosper. Ignore it and we perish. [1]
>
> \- *Harry Newton*

"The Information Age" has arrived. The advanced industrial countries are experiencing an explosion of new information technologies, services, and data processing capabilities. Technological innovations in computers, microelectronics, and telecommunications are fundamentally altering the way people of the so-called "post-industrial society" create, store, process, and use information. This is a study of the complex political issues surrounding one such information technology, satellite communications, and how that technology affects the relations between countries.

Information Competence

Information technologies pose some fundamental political questions: How do information technologies change the military, economic, and political relationships between countries in the international system? Do they create a new source of national or global power, or merely reinforce those powers already in existence? The central focus of this book is the relationship between information technologies and national power.

The most basic question in this investigation became: does technology "lead" politics, or does politics "lead" technology? In order to guess at the direction of causality the researcher must closely examine both the technology and the politics. The emphasis on the "technological perspective" in this study results from observing the differing rates of change of information technologies and politics [2]. This is not to say that political, and especially national security, factors are usually secondary when analyzed in conjunction with information technologies. Political factors are historically the prime movers and chief motivations for the massive governmental financial investments that are required for technological innovations and development. However, since the 1960s, information technologies have gained their own developmental momentum. Drawing on the innovative energies and technological infrastructures created by the massive investments during that decade for the politically motivated "space race," expanding information technologies and services increasingly blur traditional lines of governmental jurisdictions as they create new areas of human activity and interaction.

Technological innovations which create their own demand have a synergistic effect on the society that produces and uses them. The multinational corporation that uses satellite-linked teleconferencing not only saves time and money by not shuffling executives between rooms in different parts of the world, but, more importantly, the faster and more efficient exchange of information through electronic conferencing extends and enhances that firm's administrative control to any region or market on the globe.

The emergence of information technologies as major factors of economic, political, and military power portends far-reaching implications for the future structure of the global community of states in the "Information Age." It now appears that the political issues posed by information technologies in general, and satellite communications in particular, are predominately technology-driven. The new information and telecommunications technologies are creating novel capabilities and services at an extremely rapid rate — so rapid that national and international governmental entities are hard pressed to fill the "policy spaces" left in the wake of such invention. Remote-sensing satellites, mobile satellite services, and direct broadcast satellites exemplify the technologies that are floating in policy vacuums, both on the national and international policy-making levels. In short, this book investigates the policy vacuum in outer space that has been created by communication satellites.

A growing number of international political observers contend that information technologies have forced a change in the way a state's power is assessed. According to these analysts, a country's power should no longer be measured solely in terms of land area, populations, industrial output, military preparedness, or resource deposits. In the book, *The Fifth Generation*, authors Edward Feigenbaum and Pamela McCorduck, describe the pioneering efforts of the Japanese to create "Artificial Intelligence." The authors emphasize the incentives driving Japan to be in the forefront of the race to develop the first machines capable of human-like thought and point out the implications of this revolutionary innovation for the global distribution of power:

> The world is entering a new period. The wealth of countries which depended upon land, labor, and capital during its agricultural and industrial phases — depended upon natural resources, the accumulation of money, and even weaponry — will come in the future to

> depend on information, knowledge, and intelligence. This isn't to say that the traditional forms of wealth will be unimportant... But in the *control* of all these processes will reside a new form of power which will consist of facts, skills, codified experience, large amounts of easily obtained data, all accessible in fast, powerful ways to anybody who wants it — scholar, manager, policymaker, professional, or ordinary citizen. And it will be for sale. [3]

The wealth and power of a state will increasingly depend upon its people's ability to efficiently create, process, organize, and communicate information. In other words, information technologies have created a new, and perhaps, determining factor to consider in assessing national power — a society's *information competence.* All forms of national power — military, economic, and political — are directly affected by a state's information competence. This view was articulated in an article appearing in *Le Monde* describing the United States' policies regarding science and technology *vis-a-vis* Japan and the Soviet Union:

> Les technologies de l'information (informatique et communication) sont vitales pour asseoir leur suprematie tant militaire qu'industrielle. (Information technologies are essential to ensure as much their industrial as well as military supremacy.) [4]

Defining Information and Communication

In the realm of political, or more specifically, "power" relationships, information usually is of little interest unless it can be used to invoke a response or reaction in another actor or actors. I have defined information as *that which is communicated.* Communication is the transfer, exchange, and dissemination of ideas and messages through a particular medium, whether by voice, computer, or satellite.

However, in the following chapters we will delimit our definition of information to that which is communicated *electronically*, in a general sense, and that which is communicated electronically via satellite in our more specific treatment of the politics of communications satellites. Even though at first glance this appears to be a broad delimitation, satellite electronic communications encompass almost all forms of electronic communication — voice, data, print, video, audio, broadcast — in the performance of a vast range of political, economic, cultural, and military functions.

This definition of information anticipates the technology's accelerating development. Computer data processing and communications technologies and services are merging within integrated all-digital networks. Telecommunication networks based upon analog techniques were electronically separate and distinct from digitally-based computers.

The technological marriage of information and telecommunications systems made possible by digital switching equipment introduced in the late 1970s will consummate in the "integrated services digital network" (ISDN). ISDN will create a "transparent" network in which the user is unaware of distinctions between data processing and communications systems. In essence, the global telephone network is becoming a worldwide computer grid. However, ISDN is coming into a political and legal world marked by a myriad of post office, telecommunications, educational, cultural, economic, and national security ministries and agencies in the more than 158 countries whose jurisdictional boundaries still adhere to the previously distinct technological divisions.

Converging computer and telecommunications services are bringing these administrative entities into bureaucratic chaos on all levels — from the village telephone exchange to international "teleports." The fact that some groups or industries are favored or better represented in one agency more than another means that the blurring convergence of techno-jurisdictional lines instigates much of the political divergence we see today as countries grapple with information policy.

Communications, National Power, and Information Competence

How is information power? Information reduces uncertainty. A policymaker with access to a greater amount of information has the advantage of a more complete knowledge of a situation's power factors and available action options. Communications means moving information from its collection to its processing and then to its use. Communications systems play crucial roles for the timely implementation of policy decisions.

Information and communications technologies are often called *force multipliers*. They promote the most efficient use of resources — both military and agricultural. The country with superior information competence will have an advantage over a country with low competence. This was shown to be the case in the 1982 Falklands War where British forces — who had access, via high-capacity satellite links, to reconaissance and weather satellite data, as well as to the combined expertise of the headquarters staff 8,000 miles away — decisively defeated the Argentines — who had difficulties just maintaining radio contact with their forces only 400 miles offshore. The Falklands example illustrates that the politics of communications are the politics of access to these technologies.

Access

Access to advanced information and communication technologies (sometimes called "informatics" or "telematics") is possessed by very few countries. As such, access to, or the ability to use these technologies can be considered a scarce resource — a resource with far-reaching power consequences. For example, the informatics infrastructures of modern post-industrial countries such as the United States, Japan, and France, can relay "gigabits" (gigabit equals one billion bits (or pieces) of computer-usable information) instantaneously almost anywhere on earth to serve that country's political, economic, and strategic interests. In contrast, it can take hours, if you're lucky, to make a telephone call across the city of Lagos, Nigeria, or Cairo, Egypt.

Unequal access creates both new configurations of power and "interest groups" of countries that attempt to maintain their advantage or to redress the rules of the existing regime to better support their own national and collective interests. These are cross-cutting coalitions. For example, although the Soviet Union has a relatively advanced telecommunications infrastructure, it generates very little international telecommunications traffic [5]. Access alone does not create an "information society." It is significant that in an area as dynamic and profitable as information technologies, the highly industrialized societies of the Soviet Union and her allies are relegated to the status of fascinated onlookers. What then, is access?

Access, in practice, is a function of both the technical ability of the user to utilize the technology and the user's needs and requirements. This is a crucial consideration for the political and economic sides in the ongoing discussions and negotiations over the issue of access. Many countries would not be able to use computers were they suddenly to appear; their economic and social infrastructures are not geared to that type of information processing. Developing countries also lack the requisite telecommunications network with which the computer receives and transmits data and programs.

As the Soviet Union example shows, the ability to access international telecommunications networks does not by itself generate the need or demand to communicate internationally. The particular economic and political stance of the Soviet Union does not encourage or demand the extensive and sophisticated telecommunications required by Western multinational corporations for their international banking and other operatons. In addition, and not incidentally, the Soviet policy of hindering the movement of citizens both within and out of the country eliminates much of the source of international telephone traffic [6].

Use and Access

Paradoxically, information alone does not represent power, but the distribution of access to communications and information technologies, and the ability of the user to use the technologies does. The uneven distribution of access has created "information-rich" and "information-poor" countries. This points to two axes in the geopolitics of information: use and access.

"Use" refers to the content, direction of flow, and end effects stemming from the utilization of an information system or technology. Use is objectively political. In contrast, "access" denotes the ability to use the information technology; access is objectively technological. To complete the analytical framework, "use" corresponds to the political perspective, while an analysis of "access" requires the technological perspective. Use politicizes access, while the reverse is less often the case. However, when one side finds itself at a disadvantage or powerless in its attempts to change the ways the technologies are used, it may broaden the debate to include the access dimension where a different and possibly more favorable distribution of power may prevail.

We find this to be the case when we look at the issues surrounding satellite communications. The global politics of communications satellites is predominately the politics of access, but instead of merely being a question of access to the two outer space resources that satellite technology must use — the geostationary satellite orbit (GSO) and the radio frequency spectrum.

Briefly, the geostationary orbit (22,300 miles or 35,780 kilometers above the equator) is the only flightpath from which a satellite appears to hang motionless in the sky, allowing very reliable and long-distance telecommunications pathways over most of the earth's surface. All communications to the satellite and back to earth must use electromagnetic waves in designated radio frequency spectrum bands. Politics, or the process of determining "who gets what," enters the picture because the resources are limited. Many countries perceive that there will be too many satellites, too close together in the geostationary orbit, all trying to use the same frequency bands, resulting in a telecommunications gridlock in space. Will the newcomer seeking to place his space vehicle in the geostationary telecommunications traffic flow be blocked by those already there? While we can say that communications satellites utilizing the geostationary orbit and radio spectrum have politicized these resources, the question remains — why? How does a telecommunications technology politicize outer space resources?

This book focuses on the reasons for the politicization, the states and international organizations involved, and suggests a perspective that may illuminate practicable solutions to the problem of devising an equitable allocation of the resources. This is the challenge now being faced by the International Telecommunication Union (ITU), the United Nations specialized agency that oversees the radio spectrum and orbital resources. At two ITU World Administrative Radio Conferences in 1985 and 1988, the more than 158 countries belonging to the organization will promulgate international law allocating and regulating the resources.

The technologists' need to comprehend and appreciate the political aspects of the language, issues, and goals of the participants in the communications satellite debate is clear. While the problems of satellite congestion appear technical, the universe of possible solutions is ultimately funneled by politics — the power interests of states. We will look at the international debate over communications satellites from the two political dimensions that we have indentified as the inherent in all information technologies: use and access. The central question then becomes, are countries demanding access rights to the geostationary orbit and radio spectrum resources as a means of exerting influence over how satellite communications are used, or is it a debate over access, *per se*?

My argument is that while the debate — especially in the ITU but also in other fora — started out as one over access to soon-to-be saturated resources, technological advances are unmasking that rationale, revealing items on the hidden agenda behind the politicization of communications satellites and the orbit/spectrum resource — the so-called New World Information and Communication Order and space weaponization. To get from here to there involves traversing a fascinating labyrinth of murky political passageways illuminated by a truly miraculous technology, which, in the words of the originator of the geostationary satellite, Arthur C. Clarke, "will link together the whole human race, for better or worse, in a unity which no earlier age could have imagined." [7] To begin our trek, we now turn to the use dimension.

REFERENCES

1. Harry Newton, "Ours Are the Best of Times," *Newsweek*, October 31, 1983, special edition to Telecom '83.
2. See Henry R. Nau, *National Politics and International Technology: Nuclear Reactor Development in Western Europe*. Baltimore: Johns Hopkins University Press, 1974, pp. 9-32. Nau's first chapter is an insightful discussion of the "technological" and "political" perspectives. He quite correctly asserts that the researcher should make his or her assumptions explicit to the reader. Although Nau adopts the political perspective, he also states, " . . . these results are each valid *within the perspective adopted* [sic]." I encourage the reader to refer also to Langdon Winner's book, *Autonomous Technology: Technics Out-of-Control as a Theme in Political Thought*. Cambridge, MA: MIT Press, 1977. This is a work that analyzes the technological perspective of politics.
3. Edward A. Feigenbaum and Pamela McCorduck, *The Fifth Generation*. New York: Addison-Wesley Publishing Company, 1983, p. 14.
4. Eric Le Boucher and Jean-Michel Quatrepoint, "La Guerre Mondiale de la Communication," *Le Monde*, January 11, 1984.
5. The Soviet bloc satellite network, INTERSPUTNIK, carries only 0.3 percent of the traffic carried by INTELSAT.

6. The Soviet Union offers us a fascinating example of how the centralized leadership of a country attempts to come to grips with decentralized information and data networks. The information competence necessary for sustained economic and technological growth requires greater and greater computer resources and computer literacy in the population. This, however, threatens to undermine the government's monopoly of information and, thereby, control deemed crucial by the Party for its survival. See John Tagliabue, "Hungary Encourages Interest in Computers," *New York Times*, January 7, 1985.
7. Quoted in Ashis Gupta, "Satellite Communications: Forward to the Brave New World," *IEEE Spectrum*, September 1984, p. 9.

CHAPTER 1

USE OF OUTER SPACE FOR POWER ON EARTH: TELECOMMUNICATIONS AS A NATIONAL INTEREST

Communications is addictive. The easier and less expensive it is for people to communicate, the more they will.

- *William G. McGowan*

At any particular time an average of 20 percent of all installed telephone capacity in African urban centers is out of order. [1]

- *J.S. Malecela*

Introduction

Since the late 1960s, communications satellites have been the most important means of relaying and distributing information on a global basis. Satellites provide two-thirds of the present international telephone circuits and all "live" intercontinental television links. Access to satellite communications is, in many respects, access to worldwide information flows and technologies.

The Satellite Networks: Multimedia "Stars" in Space

Since the launching of both Early Bird, the first commercial communications satellite, and its parent organization INTELSAT in the mid-1960s, satellite communications has made possible the creation of the world's most "complex" machine — the global direct-dial telephone network. In addition, Early Bird provided the telecommunications link for the first transoceanic, transcontinental television broadcasts. In the 1980s, television is the fastest growing service on the INTELSAT network, which relayed over 51,000 hours in 1982, an increase of 26 percent over 1981 [2]. An estimated television audience of 2.5 billion people watched the 1984 Los Angeles Olympic Games "live via satellite."

Access to satellite communications and the orbit/spectrum resource is available to all states as members and participants in the global satellite networks, INMARSAT, INTERSPUTNIK, and INTELSAT. Let us now take a closer look at the three global networks.

INMARSAT and INTERSPUTNIK

INMARSAT (International Maritime Satellite Organization), is the "youngster" among global satellite networks. The organization, with headquarters in London, was established on July 16, 1979 as the result of a series of conferences convened by the Intergovernmental Maritime Consultative Organization. The system commenced full-scale operations on February 1, 1982, using leased capacity on INTELSAT and US Navy Marisat satellites. INMARSAT provides its 40 member states a wide range of telecommunications services between properly equipped maritime vessels and land-based telecommunications networks. Currently, INMARSAT utilizes leased transponders on INTELSAT V-MCS (maritime communications subsystems) satellites, Marisat satellites, and the European-built Marecs satellites.

The constant need for reliable maritime communications and safety and distress signaling has made INMARSAT a truly universal organization. Notable on its membership list is the Soviet Union, a non-member participant in the INTELSAT system. Olof Lundberg, INMARSAT's Director-General, stated that the Soviet Union joined as a full member because, with its large fleet of ships and telecommunications traffic, its share of space segment utilization would grant it a sizable vote in the INMARSAT council. "The 14 percent INMARSAT shareholding by the Soviet *Morsviazsputnik* group is second only to the US COMSAT's holding of 23 percent."

Even more significantly, thanks to its membership in INMARSAT, the Soviet Union also gains legitimate access to advanced Western satellite technology. One avenue to that technology is through the provision of launch services. The Soviet Union has proposed to sell launch services for INMARSAT satellites on their *Proton* rocket. In 1983, the Soviet representative quoted a launch cost of $24 million, a price that competes with rival European Ariane and US Space Shuttle systems. If a launch contract were granted to the Soviet Union, the technical requirements for a precise "fit" between both the Proton launch vehicle and the satellite payload, in terms of both hardware and software, would allow Soviet technicians and engineers access to the most intimate details of the satellite's technology. Some observers contend that, in addition to the hard currency earning such a venture would generate, this may explain the Soviet willingness to open up their space program to a Western-dominated satellite consortium [3].

The technical characteristics of the INMARSAT system that enable it to provide excellent quality maritime telecommunications are applicable to almost any type of mobile communications system. INMARSAT is planning to establish technical standards for aeronautical and land mobile systems, the first step toward eventual system design and implementation. Many industry observers consider mobile services as the markets of greatest potential demand for satellite communications services [4].

INTERSPUTNIK, "The *Intersputnik* Organization of Telecommunications Satellite Users," is the Soviet bloc counterpart to INTELSAT. Of the 14 signatory members of INTERSPUTNIK, seven are in Europe (Bulgaria,

Hungary, Czechoslovakia, Poland, German Democratic Republic, Romania, and the Soviet Union). Mongolia, Vietnam, South Yemen, Afghanistan, Syria, and Laos are the Asian members. Cuba is the only signatory in the Americas, although Nicaragua has joined the system as a participating member. As with INTELSAT, other states are non-member participants: Algeria, Iraq, and North Korea use the *Gorizont* satellites for telecommunications with INTERSPUTNIK members.

With 14 members and only 0.3 percent of the traffic volume carried by INTELSAT, INTERSPUTNIK at present plays a very minor role in international satellite telecommunications. However, for technical and political reasons, INTERSPUTNIK could become a potent competitor to INTELSAT in the future.

Technologically, INTERSPUTNIK utilizes transponder capacity on the powerful Soviet *Stationar* satellites requiring smaller and less expensive earth stations. Politically, INTERSPUTNIK's charter gives each country one vote, regardless of its telecommunications traffic volume of use of the space segment. In contrast, INTELSAT (as well as INMARSAT and ARABSAT) is run by a Board of Governors in which the states with the largest volume of traffic are alloted votes commensurate with their use of the space segment. Ironically, although INTERSPUTNIK was established to assert Soviet foreign policy objectives and prestige, most Soviet bloc countries are users of the higher quality INTELSAT circuits [5].

INTELSAT: The Global Connection

INTELSAT is the only truly global and universal satellite network. The only legal requirement for membership in INTELSAT is that the applying country be a member of the ITU. At the time of this writing (March 1985), INTELSAT had 109 member states and provides satellite telecommunications services to some 170 countries, territories, and areas of special sovereignty. As stated before, even the Soviet-bloc countries are non-member users of the INTELSAT satellites. The universality of the INTELSAT system in both policy and practice is emphasized by one of its publications:

> The prime objective of INTELSAT is, of course, the provision, on a commercial basis, of the space segment required for international public telecommunications services of high quality and reliability to be available on a *non-discriminatory* basis to *all areas of the world* at the lowest possible cost [emphasis added].[6]

Economically, INTELSAT has both developed and benefited from innovations in microelectronics, space systems, and information processing. The system started operations in 1965 with the Early Bird (INTELSAT I) satellite.

Early Bird boasted a capacity of 240 telephone circuits or one television channel. Today INTELSAT provides over 50,000 telephone circuits over a network of approximately 15 satellites. Costs per circuit have declined from $64,000 (circa 1965 dollars) to $4,600 (circa 1983 dollars) — about one-fifteenth of the original cost.

Starting with two earth stations in 1965, the system now spans the three ocean regions through more than 300 earth stations in over 134 countries. The INTELSAT system provides over 1600 earth station-to-earth station pathways, or more than 100 such pathways from each satellite. This is an important consideration for evaluating the efficiency of orbit/spectrum resource use, with the measuring criterion being the number of pathways supplied by a satellite in a particular orbital slot.

INTELSAT is currently developing its sixth generation satellite, INTELSAT VI, which, when propelled up to the geostationary orbit in the late 1980s, will simultaneously provide up to 40,000 telephone circuits and two color television channels. INTELSAT's technological improvements have succeeded in lowering costs, improving quality, and making telecommunications available to more countries than ever before. These improvements have also had the effect of creating demand for telecommunications services, so much so that the volume of international telecommunications traffic doubles every four to five years.

Although INTELSAT was established specifically to provide international satellite telecommunications, it is now also the largest supplier of domestic telecommunications networks and services. In 1974, Algeria proposed to INTELSAT that excess transponder capacity be used for purely domestic telecommunications purposes. INTELSAT began offering domestic transponder leases in the mid-1970s on a pre-emptible basis, a service that is among its most demanded. INTELSAT reports that in mid-1985, 26 countries are leasing some 40 transponders for domestic services, with 13 countries planning to lease or expand their use of INTELSAT's domestic services. In fact, there are more earth stations in the INTELSAT network providing domestic services (over 350 antennas in 1984) than the number of earth stations (about 100) used specifically for international traffic [7]. Significantly, of the 26 countries presently utilizing INTELSAT leased transponders for domestic telecommunications, 19 are less developed countries (LDCs).

The number of INTELSAT domestic transponder leases is strong evidence of the growing demand for satellite communications, and indicates where that demand is most keenly felt — in the LDCs. This is all the more remarkable given the relative expense of INTELSAT earth stations and pre-emptible transponder leases to countries seeking to build a domestic telecommunications infrastructure. However, INTELSAT argues that it nonetheless provides states with the most efficient and economical access to the orbit/spectrum resource for both international and domestic satellite communications [8].

1. Global Cooperation for Profit

The INTELSAT system works, but how does it function? Why do countries belong to and participate in the system? And perhaps more important politically, does INTELSAT provide the "equitable access to the orbit/spectrum resources" demanded by many LDCs?

Briefly, INTELSAT is the owner and operator of the space segment, composed of the satellites, tracking and telemetry control facilities, and the administrative machinery coodinating procurements, planning, and financial management. Member countries, in turn, own a portion of the space segment commensurate with their investment share and traffic volume. They are billed for this traffic and at the same time receive dividends proportional to their investment share.

The earth segment — antennas, receivers, transmitters, and other ancilliary equipment — is owned by the member country or designated entity. Tariffs actually charged to customers are determined solely by the member country; INTELSAT charges remain the same regardless of what is charged the end user. For example, in 1980, the Federal Republic of Germany paid INTELSAT $9.7 million for the use of INTELSAT's circuits. During the same period of time, after subtracting overhead costs, the West German Bundespost made a *profit* of over $40 million [9].

INTELSAT is even a money maker for developing countries with low per capita disposable incomes and few telephones. In many cases, countries can actually receive more revenues in return — in the form of INTELSAT dividends—when they charge high tariffs to their customers. The high tariffs have the effect of reducing those countries' volume of traffic on the INTELSAT system and increasing domestic revenues, while on the global level, the expanding volume of INTELSAT network traffic increases total system revenues which these states receive as dividends. As a result, INTELSAT is often seen as a revenue producer by policy-makers in many countries that are more concerned with short-term cash flow and less aware, perhaps, of the long-term financial and societal benefits that inexpensive and universal telecommunications services offer national development. David Leive, General Counsel for INTELSAT, estimates that the space segment tariffs charged by INTELSAT account for only about 10 percent of the total amount charged to customers worldwide [10].

However, INTELSAT's claim to provide the most economical and efficient access to and use of the orbit/spectrum resource is seen by some observers as a cost/benefit ratio that varies from country to country. In their view, there are two areas where one could argue that the INTELSAT ownership and technical arrangements favor big users over small. The first concerns the fact that the member country must own the INTELSAT-approved "dish" antennas and earth stations. Until recently, INTELSAT had allowed only two antenna sizes to be utilized in the space segment: the 30-meter diameter "Standard A" and the 10-meter "Standard B." For many LDCs with few telephone subscribers, the cost of the earth station and antenna is, relative to their amount of traffic and system dividends, proportionately higher than for highly developed countries (HDCs) with greater traffic volumes. This is due to the fact that the fixed costs of using INTELSAT are concentrated in antenna and earth station equipment costs. For example, the country with only 100 circuits on the INTELSAT satellite must pay the same earth stations costs as the country with 1000 circuits, the latter receiving ten-fold the amount of dividends as the former.

This ownership arrangement is being attacked from two directions by proponents of less-expensive satellite communications for LDCs. First, many LDCs are penalized for being low volume users of the space segment. Countries using the smaller and less expensive Standard B antennas are charged 50 percent more per circuit due to the antennas' "less efficient use of the space segment resources" [11].

Secondly, INTELSAT critics contend that the performance standards required for INTELSAT certification make the antennas too expensive for the financially-strapped Third World countries. In 1983, Professor Bruce B. Lusignan, Director of the Stanford University Communications Satellite Planning Center, pointed out that equipment standards established and promulgated by the International Telecommunication Union and INTELSAT are not appropriate for the development of communication infrastructures in many developing countries. The excessively restrictive standards raise the cost of satellite telecommunications equipment two to five times the price of similar hardware available in deregulated US markets. According to Lusignan,

> The problem is that large US companies like Motorola and GTE dominate US preparations for the [international] conferences that establish international standards and benefit from those standards when it comes to marketing their equipment internationally. Moreover, agencies such as the Federal Communications Commission, are "two-faced" when it comes to standards. While the FCC moves to open equipment markets to competition in the US through loosening technical standards, it often seeks to develop costly and inflexible equipment standards internationally. Small companies cannot afford to represent their interests in international organizations such as the ITU [while] US companies contribute 3 percent of the ITU's annual budget.[12]

Curiously, the technologically advanced United States manufacturers may be creating their own competition. These firms, striving to maintain their market share through the imposition of high technical standards for INTELSAT and the other regional telecommunications systems intended to meet the "low-technology" needs of the LDCs, are, in effect, encouraging the establishment of cheaper competing satellite systems for INTELSAT. "Low-tech" systems could provide a wide range of domestic and regional telecommunications services with a level of technology and performance that is both functionally and financially "appropriate" for LDC needs and resources.

In support of his contentions, Lusignan cites the work being done between the Communications Satellite Planning Center at Stanford University and the cross-town firm, Equatorial Communications of Mountain View, California. Both of these Silicon Valley organizations provided the Rome-based Intergovernmental Bureau of Informatics with small earth stations in a demonstration of data networks at the Geneva Telecom '83 exhibition. The stations used .6 meter antennas and spread spectrum techniques to receive data over INTELSAT transponders at a total cost of $2,000 per station. Using slightly larger 1.3 meter antennas, earth stations capable of also transmitting data (not voice) on existing INTELSAT transponders would cost approximately $6,000. The potential for this technology to expand data networks is enormous. As Lusignan states, "We are the guerilla force in satellite communications—making a wide sweep around the established powers. The New World Information Order is achievable today with present technology" [13].

2. Equitable Use?

INTELSAT as *the* major provider of global satellite telecommunications is also *the* means of access to the orbit/spectrum resource for its member and non-member participating countries. As previously stated, INTELSAT is owned by the member countries who use it. A member's investment share equals its "percentage of all utilization of the INTELSAT space segment by all Signatories." According to Eilene Galloway, this arrangment is "an example of *equitable* sharing which is legally different from the idea of *equal* sharing" [emphasis added] [14]. In line with Galloway's assessment, it would appear that INTELSAT does indeed provide all members with equitable access to the orbit/spectrum resource and satellite communications commensurate with their *use* of the system. This is a point of controversy in the debate over communications satellites. The LDCs wish to translate "use" to "need" for the system in assessments of equitable access being negotiated in the ITU, INTELSAT, and other fora. In other words, does INTELSAT provide the type of satellite communications the LDCs need? Not unexpectedly, the controversy has technological as well as political and economic dimensions.

Critics point out that INTELSAT was formed by the United States and other HDCs to provide satellite telecommunications corresponding to their needs in the 1960s and 1970s, not necessarily to those of the Third World in the 1980s and beyond. INTELSAT's technical standards and network architecture were established for intercontinental telecommunications, not for thin-route rural and low volume applications. Given the relatively high fixed earth station costs for low volume LDC users, "equitable access" comes at a higher price per circuit. The higher prices are found in the utilization charges assessed to users of Standard B antennas, the high technical standards for INTELSAT-approved equipment, and finally, the requirement (now being re-negotiated) that a country wishing to lease transponder space on a *non-preemptible* basis must pay twice as much than for a pre-emptible transponder [15].

3. INTELSAT's Response

INTELSAT's technological and economic perspectives are changing in recognition of the large market and user demands in LDCs and especially to the challenges posed by competing satellite systems, such as ARABSAT, EUTELSAT, and private North Atlantic systems led by the Orion system's proposal. INTELSAT realizes that the vast, virtually untapped market for inexpensive satellite telecommunications systems in the LDCs, made possible with high-power satellites, is a service it must provide if it is to survive as an organization into the 21st century [16].

In late 1983, INTELSAT began its offensive against the competing systems vying to cut vital pieces from its market. In December of that year, INTELSAT introduced its "VISTA" service. VISTA would cater to the low-volume domestic and regional telecommunications needs of many developing countries and international organizations located at remote development sites. These networks would use a new "Standard D" earth station costing in the neighborhood of $25,000 to $50,000.

Borrowing from its experience with the IBI and Equatorial Communications data transmission experiments, INTELSAT announced in September 1984 that it would implement the "INTELNET" service. Data terminals would consist of a personal computer connected to a satellite receiver and a 65 centimeter (2 foot) antenna. INTELNET could disseminate data anywhere within the global beam. For example, UN development advisors could carry terminals with them to remote villages and projects. Although small antennas and low-powered INTELSAT satellites make voice transmission with INTELNET impossible, the system nonetheless moves the global consortium closer to providing the telecommunications service envisioned by the New World Information Order. However, declining costs encourage countries to establish satellite systems, many of which will directly compete with INTELSAT.

4. Regional Satellite Systems

Since the first domestic geostationary satellite network was established by Canada in 1972, the number of states either building or planning regional systems has steadily grown. The two most notable are the ARABSAT and EUTELSAT systems [17]. The ARABSAT Satellite Communications Organization was established through an agreement signed by the 22 members (including the Palestine Liberation Organization) of the Arab League in 1976. Two ARABSAT satellites will provide telephone, data, television, and radio services to earth stations located throughout the Middle East and Northern Africa. Aerospatiale of France is the main contractor of the project, while Ford Aerospace is the subcontractor responsible for the satellites' communications payloads. A look at the objectives of the ARABSAT system indicate that it is being implemented for other than solely technical or economic reasons.

According to the organization, the priorities of the ARABSAT system are:

> 1. The establishment of reliable communications links between the Arab States.
>
> 2. Communications networks in rural areas, especially for programs concerning education, information, and culture.
>
> 3. Program exchanges.
>
> [The] ARABSAT Organization shoulders the responsibility for the transfer of technology in the field of space technology and is bent on the promotion of space communications industries in the Arab World and on the conduct of studies and research in space technology.[18]

The EUTELSAT System (European Telecommunications Satellete Organization) was established in a provisional form on June 30, 1977 by the 17 members of the European Conference of the Postal and Telecommunications Administrations (CEPT). The definitive form of EUTELSAT was finally agreed upon by 26 European countries on May 14, 1982, in the adopting of a permanent convention replacing the provisional agreement. Under the Convention, the European Space Agency (ESA) will "provide, launch, and position in orbit, the ECS [European Communications Satellite] using the European launcher Ariane." [19] The EUTELSAT system pursues technical and economic objectives in serving as a vehicle for pooling European technological and financial resources in the expansion of European research and development.

In contrast, the ARABSAT system clearly reflects the political motivations behind regional and national satellite networks [20]. The ARABSAT satellite system is a groundbreaking cooperative venture among Arab countries. As a regional satellite system, it will provide telecommunications and direct broadcast television to the countries in its coverage area. Most importantly, it offers these countries access to high technology and reduces their perceived dependency on INTELSAT's provision of domestic and regional services. Furthermore, ARABSAT is a very high prestige project for Arab countries, officially marking their entry into the Space Age. The following statement by ARABSAT's Director-General, Ali Al-Mashat, summarizes the technological, economic, and political incentives pushing an increasing proliferation of dedicated national and regional LDC satellite systems:

> No doubt, the realization of the ARABSAT System and the attainment of its proclaimed objectives shall serve the economic and technical independence of the Arab region, *augment the sense of national prestige*, play an important role in the organization of economic and social structures by strengthening the linguistic and cultural homogeneity, by facilitating decentralization of industry, by slowing the migration of population to urban centres, by maximizing the use of available centralized facilities and expertise, etc., and lead, finally, to the realization of the Arab's aspirations in prosperity and unity [emphasis added].[21]

Political incentives of national, regional, and cultural prestige, as well as of enhanced control over the flow of information between states and regions are instrumental in propelling states to commit scarce resources for satellite systems apart from INTELSAT. While the establishment of these satellite systems is rational from each individual country's point of view, the consequence of such actions for the community of orbit/spectrum users is an "irrational" crowding of the resource. It is "irrational" since the orbit/spectrum resource is degraded for all users, even those establishing satellite systems. Whatever the cause, the end effect is clear — increased crowding and conflict over the allocation of rights to the orbit/spectrum resource and increasing politicization of satellite communications.

Politics of a Place in Space

We will now focus our investigation onto the politicization of the orbit/spectrum resource; why are they becoming increasingly salient as sources of international competition and, potentially, strategic conflict? This question raises a plethora of ancilliary issues as wide-ranging as the uses to which satellites are applied — from television broadcast to military communications. Is the debate over the orbit/spectrum resource really a controversy over *access* to the resources, or is the issue of *use* of the orbit/spectrum resource a more prominent concern to policymakers, one that influences their stance toward the question of access? There are many factors indicating that a state's perception of *use* (i.e., what communications satellites do in space and whether or not their use of the resources is in a particular country's national interest) in fact determines that state's policy stance towards the issue of *access* to the orbit/spectrum resource and satellite communications.

We will now examine the use dimension of satellite communications. The following discussion centers on two points: (1) how and to what purposes the orbit/spectrum resource and satellite communications are used; and (2) the political implications and consequences of that use for countries' policies regarding the issue of *access* to the orbit/spectrum resource.

The Power Dynamics

To the extent that satellite communications enhance the user's ability to create, store, process, and relay information (i.e., "information competence"), countries see definite national interests at stake regarding how the orbit/spectrum resource is allocated and used. As one study found:

> How the world community manages the growing demand and competition for access and use of the orbit/spectrum resource will affect the quantity, distribution, quality, and price of information transmitted electronically.[22]

The debate over the use of the orbit/spectrum resource and satellite communications is polarized predominately along "North-South" lines —"North" referring to the HDCs and "South" referring to the LDCs. Many less developed countries see the manner in which the orbit/spectrum resource is currently used as widening the communications and information "gap" between themselves and the highly-developed countries. These LDC concerns were expressed in the New International Ecomonic Order and New World Information and Communication Order (NWICO) [23].

However, the issues also have an intriguing East-West dimension as well. Here we return to the re-definition of power mentioned in the Introduction. The importance of information technologies for many aspects of national power compels a reassessment of the relative power status of the Soviet bloc states *versus* the Western post-industrial information societies. One of the more surprising aspects arising from categorizing the world's countries between the information "haves" and the information "have-nots" is that although the Soviet bloc states are heavily industrialized, they are not "information societies." Ranked in terms of their information competence, the Soviet Union and other Warsaw Pact states are, in reality, less-developed countries [24]. The Soviet bloc's ambiguous stage of development between post-industrial and information societies can be seen in ambivalent policy stances toward NWICO issues and negotiations on the access to the orbit/spectrum resource. However, in the broad perspective, the Soviet and LDC policy stances toward use correspond to what we will designate here as the NWICO view.

In this view, the current use regime favors the few countries who enjoy the benefits of national (and especially military) satellite sytems, without the constraints on use and dependency on foreign satellites operators associated with common user systems such as INTELSAT [25]. As Anthony Smith writes:

> The trouble is that the high technology of today tends to make societies dependent upon Western suppliers and often upon satellites. The satellite enables developing societies to link themselves together at a tiny fraction of the cost in money and labor. But it means that more and more functions within the linked societies become dependent upon a foreign-built satellite.

Dependency costs link the issues of use and access together. On the use side, HDCs are dependent upon the cooperative-use regime for orbit/spectrum resource allocation, while LDCs depend on foreign owned and controlled satellite networks for access. The accelerating transformation of HDCs to information-based economies and societies is dependent on their ability to communicate data efficiently and cheaply, a capability requiring ever growing amounts of spectrum and orbital slots. Ironically, even as HDCs realize the power advantages offered by high capacity satellite global networks for multinational corporations or military alliances, they are also cognizant of their growing dependency on orbit/spectrum allocations as well as their vulnerability to spectrum coercion. LDCs, realizing this, seek to exploit HDC access vulnerability for bargaining leverage in negotiations within the NWICO, as well as in global economic policy negotiating fora as part of their call for a redistribution of world wealth. As Ronald Stowe, General Counsel for Satellite Business Systems writes:

> We [the United States] are confronted in the 1985 WARC with a serious, broadly-based, and unavoidable attempt to use the ITU's procedures and machinery for purposes which will have much more to do with economic and political ambitions than they do with technical or operating efficiency. It is quite commonly suggested by spokesmen from at least several developing countries that they should have the right through lease, sale, or barter to gain revenue from the use by others of the geostationary orbit.[26]

This is a central, but non-discussed aspect of the access debate in the ITU. Are the LDCs really desiring access, *per se*, or are they using the access issues as a means to pursue other policy goals regarding international information policies? If so, it may be because the LDCs perceive a bargaining advantage derived by exploiting vulnerabilities and dependency of the HDCs on cooperative access to the orbit/spectrum resource as established by ITU World Administrative Radio Conferences (WARCs).

Dependency and Vulnerability

Satellite Telecommunications is a double-edged sword. No one country possesses hegemonic power or capabilities in either the use or access policy areas. Every form of use imposes costs due to the constraints posed by the distribution of power between the users of the orbit/spectrum resource. On one hand, the LDCs are dependent upon foreign suppliers, such as INTELSAT, for satellite provision of domestic telecommunications; on the other hand, the HDCs depend increasingly on cooperative management of the resources and on international arrangements ensuring the free flow of information.

We see a tension between the international collaboration required for efficient communications and use of the orbit/spectrum resource and the instincts of states to maximize their sovereign prerogatives by reducing their vulnerability to factors outside their direct control. On the access side, the historical success of the ITU regime is due to the "reciprocity-retaliation sanction which is an inherent part of spectrum utilization." This creates an immediate and universal interdependency compelling coordination and cooperation between users [27].

Mutual interdependencies compel cooperation, and this cooperation or policy collaboration regarding access rights is resource-driven. The alternative to a cooperative-use regime is a "war of all against all," where, paradoxically, the major users (HDCs) are also the more vulnerable to "spectrum coercion" or jamming. Communications satellites enhance power while being themselves vulnerable to all forms of radio interference, whether intentional or not.

This is one of the most important factors contributing to the politicization of access to the orbit/spectrum resource. Mutual vulnerabilities and interdependencies allow the LDCs to exploit the orbit/spectrum resource question in the ITU and other international fora as a bargaining chip for other telecommunications, economic, and strategic issues.

Noted communications economist Dallas Smythe, paints for us one possible scenario of spectrum warfare:

> Consider the following scenario: [A] dictator is toppled in a peripheral Third World country, e.g., in Central America. The United States uses economic boycott, mass media content, perhaps military aid or even open intervention. Third World countries through their own collective organizations demand that the United States stop intervention. Third World countries disrupt American use of the radio spectrum in strategically calculated steps: perhaps jamming military-used frequencies, communications satellite communication, etc. [28]

Skeptical readers who doubt the realism of Smythe's jamming scenario must look only as far as the United States National Association of Broadcasters (NAB) for evidence of real concern in the US broadcast industry. The issue concerns "Radio Marti," a Reagan Administration project to transmit US compiled news and programming to the Cuban people. It has been a point of contention since its funding by Congress in October 1983. Radio Marti began operations on May 20, 1985 and Cuban Premier Fidel Castro has threatened to jam "half the (AM) radio stations in America." Finding themselves in a real dilemma, communist Castro and the capitalist NAB are posed on the same side in their opposition to Radio Marti. Significantly, Radio Marti shows how aware the American telecommunications sectors are of their vulnerability to spectrum coercion from a "low-tech" country such as Cuba [29].

In sum, as countries evolve into information societies, we will see greater and greater dependence on cooperative spectrum use. Paradoxically though, due to the uneven diffusion of technological innovation and spasmodic rates of national development, we can expect more frequent threats of spectrum coercion as countries and terrorist groups seek to exploit opportune vulnerabilities for jamming of key national sectors. At the 1982 ITU Plenipotentiary Conference, the Soviet Union, acting through Czechoslovakia, proposed the adoption of a "national right to jam" unacceptable telecommunications.

The US and other HDCs are aware of their vulnerability. As one sales representative of a major US satellite supplier (who wished to remain anonymous) mentioned at Telecom '83, "If I had one good jammer I could put six major US industrial firms out of business tomorrow."[30] Although advances in modulation and transmission techniques (such as spread spectrum, scanning spot beams, *et cetera*) will reduce satellite vulnerability to jamming, "hardening" incurs unavoidable financial and communications capacity costs.

Let us now take a closer look at how the HDCs use satellite communications, and then shift our view to encompass the LDC perspective.

View from the North

Deterrence and Defense

Satellite communications are an integral part of national and alliance deterrence and defense systems for the Superpowers and their alliance partners. Geostationary satellites perform many missions vital to a credible and effective deterrence and defense strategy. They enhance the ability to detect threats and actual attacks at an earlier, more advantageous, point in time. Geostationary and low-orbit spacecraft can see large portions of the globe and communicate findings instantaneously to command centers. Military communications satellites in the geostationary and other orbits are an integral part of the defense establishment's "C^3" (command, control, and communications) infrastructures [31].

"C^3" is a central and crucial element of deterrence and defense strategies. Credible deterrence requires fail-safe communications to all strategic forces; communications that can "ride out" a pre-emptive nuclear strike and still remain operational and able to mount a retaliatory attack.

The orbit/spectrum resource is becoming the new "high ground" for superpower global strategic posturing. Due to their importance in strategic planning, the satellites' vulnerability to attack accelerates the militarization of outer space and the politicization of the geostationary orbit.

The need for redundant backup systems, which are capable of surviving an enemy attack, compels the military establishments of the United States and Soviet Union to deploy "spare" satellites, "harden" satellite electronics from the effects of nuclear blasts, and begin the design and construction of space battle stations that can guard themselves and other satellites. Analogous to sea-based projectors of power, cummunication satellites are becoming the "aircraft carriers" of space, both in terms of their ability to spearhead attacks and their vulnerability. It is likely that the Reagan Adminstration's place in history will be noted by future historians as the turning point when outer space, and particularly, the geostationary orbit, became battlefields for the protection of communication and other satellites.

Military Use of the Geostationary Orbit

At the present time, according to published unclassified sources, the US military establishment "relies on more than 40 satellites for long-range communications, attack warning, intelligence gathering, navigation, weather forecasts and mapping." The role of satellite communications in all areas of the military establishment is emphasized by General James V. Hartinger, Commander of the US Air Force Space Command in Colorado Springs, who states, "Over 70 percent of our long haul communications are handled by satellites." [32]

In the early 1960s, the US military quickly realized the potential of communications satellites to provide real-time communications anywhere on the globe. It achieved this in 1967 when the Defense Satellite Communications System (DSCS: pronounced "discus") fired up its operations with a worldwide network of US military communications satellites; its global dimensions indicating the far-flung needs of United States strategic and alliance commitments.

The Soviet Union in the 1960s, in contrast, focused its satellite communications program on its domestic needs to link settlements and cities and to consolidate its military command structures. Since then, the Soviet Union has deployed several satellite communications systems for both civilian and military missions. The first Soviet communications satellites were the non-stationary "Molniya" series. These satellites, placed in highly elliptical orbits, serve the extreme northern latitudes of Soviet territory, which geostationary satellites could not adequately illuminate from their position over the equator. The "Stationar" satellite series, which began in the early 1970s, were the Soviet Union's first geostationary communications satellites. Their introduction indicated the steady evolution of Soviet foreign policies and technological advances as the Eurasian superpower began to "look outward" beyond its post-World War II sphere of influence [33].

1. Satellite Communications and Projections of Power

Satellite communications project national military power by allowing widely dispersed armed forces to remain in constant communication with command centers and with each other. The ability to gather and disseminate data from all means of surveillance promotes the most efficient use of available military forces and hardware. In this way, satellite communications serve as a "force multiplier" providing forces almost anywhere on the globe with instant and continuous telecommunications command and reconaissance links. James B. Schultz comments:

> The military is in space because it makes economic and practical sense and because it provides a marked increase in force effectiveness. It is critically important that the military provides a *force multiplier effect* that permits it to do more with less.[34]

Schultz believes that the capabilities of geostationary communications satellites are particularly well suited to military needs, and continues:

> The hovering characteristic of a satellite at geo[stationary] altitude is the optimal position for our warning and communications satellites, the orbit from which we will continue to meet global military information needs. It is the high ground in space.[35]

Richard C. Henry, Commander of the Space Division of the United States Air Force, holds the following view:

> [T]he true high ground of space [is] the geostationary orbit. In this unique orbit spacecraft remain fixed over the land below; the orbit is a mountaintop relative to the earth below it. It is ideal for satellites that relay communications and provide surveillance. *Satellites in space have dramatically improved the capacity to communicate with, and control military forces* [emphasis added].[36]

Communications satellites provide "strategic connectivity." In contrast to conventional cable or terrestrial microwave relay systems, if one ground segment of the satellite system fails, the entire network does not "crash." Each earth station operates independently of the others. In addition, satellites offer flexibility of operation between users. Mobile communicators using voice modes from helicopters and airplanes can communicate with fixed command centers or other mobile units. The "broadcast" ability of satellites which allows for integrated and coordinated responses "in real time," combined with system configurations which place several satellites in the communicators' views at all times, enhances reliability and survivability of communications.

In terms of their specific missions for defense and deterrence strategies, satellites would be the first to detect the launch of missiles and aircraft beyond the range of land-based or airborne radar. The United States "Vela" satellites in the geostationary orbit scan the earth's disk with infrared sensors designed to recognize the heat "signatures" generated by rocket and aircraft exhaust as well as nuclear explosions.

The United States Navstar Global Positioning System (GPS) will use 18 satellites in various orbits to permit "users to obtain their precise locations and the time [and to] learn exactly how fast they are moving, anytime, in any weather, anywhere on earth or above it." [37] The Navstar system is designed also for use by unmanned missiles. Cruise and ballistic missiles will benefit from the navigational accuracy possible with satellites. Sources who wish to remain anonymous stated that the system can locate the positional coordinates of an object with a precision of 10 feet, in any direction. Most importantly, the system accomplishes this task for an unlimited number of users *who do not reveal themselves* while receiving the data. This is a crucial survival factor for submarines which must periodically update their inertial guidance systems, while avoiding detection.

The role played by space satellites as part of overall defense planning and tactics is outlined by the following:

> [Military] dependence on space is likely to increase. A major problem, for instance, is the defense of an aircraft carrier or other task force against attacking forces with modern munitions. Space assets in direct, near real-time communications with ground terminals at the task force can significantly extend the range at which detection, identification, and tracking can take place, particularly if these functions are integrated into suitable weapons delivery systems. Modern communication techniques can also help by improving communication security, a major consideration for United States and allied forces routinely deployed at great distances from home.[38]

Satellite communications are indispensable for major powers with global strategic and defense commitments, or for regional powers seeking to extend their influence. Satellite links facilitated projections of United States power in the 1975 "Mayaguez incident" in Cambodia and the 1980 attempt to rescue the American hostages in Iran. In each case United States forces in the theater of operations had instantaneous and constant communications with mission commanders in the region and in Washington. The effectiveness of a Rapid Deployment Force depends upon the attack units having effective global communications wherever they are sent.

The growing dependency of the United States and other countries' armed forces on satellite communications inevitably raises the question of satellites' vulnerability to enemy attacks. The following section briefly outlines the emerging threat to geostationary satellites in the high ground [39].

2. Are the High Birds "Sitting Ducks"?

The threat to satellite systems is multi-dimensional, reflecting the complexity of the technologies, the threats, and the extremes of the earth and space environments. The first dimension is the physical integrity of the spacecraft. A satellite may be directly attacked by an anti-satellite weapon (ASAT). There are three types of ASATs: (1) directed-energy weapons, (2) orbital interceptors, and (3) space mines [40].

2.1. Threat from Lasers or Particle Beams

Ground based directed-energy weapons use high-energy laser or particle beams to disable sensitive electronics or to explode propellents. At the present time, the altitude of geostationary satellites place them outside the range of ground-based energy weapons with currently achieveable power levels. However, space-based laser and directed-energy weapons operating outside the atmosphere do not have to overcome the atmospheric diffusion effects. Experts predict that these types of weapons in earth orbit will pose threats to geostationary satellites by 1990. However, the prospect of supplying the necessary electrical power for a laser weapon in outer space is presently more theoretical than practicable. There are easier ways to disable a geostationary "sitting duck." [41]

2.2. Orbital Interceptors

"Orbital interceptor" ASAT weapons are essentially orbiting conventional or nuclear warheads. Orbiting "killer satellites" are directed to maneuver alongside the target satellite and explode, thereby destroying the victim satellite. "Direct ascent" ASATs, currently being tested by both the United States and the Soviet Union, are rockets launched from a high altitude airplane. The rocket flies directly to the satellite without entering orbit. In order to defend against these types of weapons, a satellite must be able to either quickly maneuver itself out of the way once an impending attack is sensed, or to destroy the attacker with its own weapons (such as lasers).

A different type of orbital interceptor does not even have to approach the target satellite. Instead, a high altitude rocket or aircraft detonates a nuclear explosion producing an "electromagnetic pulse" (EMP).

2.3. EMP: Thinkable Nuclear War

To fulfill their missions, military communications satellites must continue to operate long after the initiation of a nuclear exchange. The chief threat to military communications satellites, other than outright physical destruction, is the electromagnetic pulse (EMP) produced by a nuclear explosion. EMP survivability has emerged as the functional criterion for communications satellites' strategic role during a nuclear conflict.

Military satellites are now designed with "hardened" electronics that can withstand an EMP "zap" and continue operating. At the time of this writing (1985), however, the EMP survivability of communications satellites in the geostationary orbit has been declared "uncertain" by experts in space warfare.

2.4. What is EMP?

EMP is a shock wave of high-energy radiation released by nuclear fission or fusion that spreads at the speed of light from the point of the detonation. The radiation produces a variety of effects depending on whether the affected object is in the vacuum of space or in the earth's atmosphere.

> When these radiations strike a metal object in space, such as a satellite, they knock out compton electrons, creating a charge imbalance in the skin of the satellite and setting up extremely high electric fields — on the order of 100,000 to 1 million volts per meter. These surface fields induce currents and voltages in the electronic payload, causing disruptions and burnouts. It is as if the delicate semiconductors that lie at the heart of a satellite were suddenly hit by a bolt of lightning. [42]

The "fatal flaw" in using nuclear weapons as ASATs is that EMP does not discriminate between hostile and friendly satellites. Military satellites are "hardened" against EMP, but whether or not it will prove effective cannot be tested fully without atmospheric atomic tests. These, however, were prohibited for ratifiers of the Test Ban Treaty (i.e., the United States and Soviet Union). Superpower awareness of EMP is a major reason why nuclear-tipped anti-ballistic missile (ABM) interceptor systems were, in effect, negotiated out of existence by the United States and Soviet Union in the 1972 ABM Treaty. Both superpowers realized the self-defeating threat to their own satellites if they attempted to use ABM systems to disable incoming ballistic missiles.

Furthermore, experts theorize that EMP poses a similar threat to ground-based electronic components with an even more far-reaching effect on society as a whole. The extensive network of civilian power grids, computers, and communications networks — actually any device or system whose functioning is controlled by microchips — is endangered by a nuclear blast hundreds or thousands of miles away in the atmosphere. As a Swiss study stated, "one blast could send an electronically advanced society back into the middle ages."[43]

Is this not the "thinkable" nuclear war? One nuclear device ignited high enough in the atmosphere so that no one on the ground was injured, nonetheless emits a pulse of radio energy that peaks thousands of times faster than a lightning bolt. In the scientists' scenario, most microchip-based devices in the information society would suffer fatal burn-outs as the wave hit the metal skin of a car, the battery wire of a pacemaker, or the electrical power line to a computer controlling a nuclear power plant. As portrayed in the somewhat melodramatic television movie, "The Day After," even cars on the freeway stopped dead after the flash because their electronic ignitions ceased functioning.

In sum, EMP poses a direct threat not only to geostationary satellites, but also to earth station networks and communications systems as a whole. However, EMP may be, in the last instance, the only way an attacker could defeat proliferating numbers of military communications and surveillance satellites. Paradoxically, EMP may be the *most usable* tactic for lesser-powers without space-based communications or advanced microelectronic-based defense systems. By exploding a high altitude nuclear weapon, one such country can neutralize much of a superpower's strategic and support assets with minimal effect on its own, perhaps less "advanced" but also less vulnerable, systems. "It is one of the defensive systems that would hurt the enemy without necessarily hurting his own people." [44]

2.5. Space Mines

The most feasible and probable threat to geostationary satellites is the so-called "space mine." One possible scenario, according to Robert Giffen, would be:

> After launch, a space mine stays dormant in the vicinity of the target satellite (within 100 km). When the time to attack comes, the mine is switched on, locks onto the target satellite, maneuvers into lethal range, and explodes its conventional or nuclear charge. Satellites at geosynchronous altitude are particularly vulnerable to space mines. [45]

3. The Software-Jamming Dimension

Communication satellites' vulnerability also has an electromagnetic dimension. Radio waves can be used to (1) "spoof", or (2) "jam" the satellite so that it is no longer controllable or that communications cannot be maintained on a reliable basis. Although lower in physical violence to satellite hardware, electromagnetic warfare can just as effectively knock out the space segment of a communications system as a direct physical or EMP attack.

3.1. Spoofing or "Space Hackers"

"Spoofing" is a technique for disrupting normal satellite functions through outside manipulation of its own systems and telemetry. This, however, is quite difficult because the attacker must know the computer command codes with which the tracking and control center directs the operations of the satellites in advance. By sending its own set of instructions, the "space hacker" can move the satellite out of its intended orbit, cause it to send information at the wrong time, or exhaust its station-keeping fuel. Military satellites are designed to operate without ground control during hostilities, making spoofing difficult. However, as with jamming, spoofing could be employed successfully against the less secure civilian communications satellites as a coercive or retaliatory action.

Concern for possible spoofing during any regional conflict prompted the designers of the ARABSAT satellite to encrypt the uplink telemetry system to prevent either Israel or another (possibly Arab) country from wrestling control from the satellites' owners [46]. One ARABSAT official emphasized that encryption was necessary also for the community reception service due to disagreements over program content:

> Encrypting spacecraft command and control transmissions is to protect the spacecraft from harm. We are looking at coding the television broadcasts because of potential repercussions to ARABSAT programs sent by one and received by another. Some Arab League members don't want a program transmitted by one country to be received by all of the others. [47]

3.2. Jamming

The intentional generation of electromagnetic interference, known as "jamming," "saturates the airwaves with electronic noise at the same bandwidth the enemy is using to communicate." The receiver antenna cannot discriminate between the "noise" and the desired signal and is unable to extract information from the radio waves. At higher frequencies, however, the antenna beams become progressively narrower, "forcing the jammer to move closer to the receiver or transmitter." In order to escape detection, a jammer could operate out of a mobile non-descript land vehicle or ocean trawler. The jammer can introduce electronic noise and disrupt communications as long as he stayed within the satellite's receiving antenna's service area. Since service areas tend to be quite large for satellites performing connectivity functions, physical proximity to the operating country is not always a major constraint for a jammer [48].

Satellite systems can be made jam-proof, but at a high cost financially and in terms of performance. Spot beam antenna technology and spread spectrum modulation techniques vastly complicate the task of the jammer (explained more fully in Chapter 2). However, these anti-jamming measures raise the cost of the satellite communications system and reduce its data handling speed and efficiency. Non-military satellites are the most vulnerable. Designed to operate for profit, commercial communications satellites lack resistance or "hardness" to jamming. This is where the vulnerability of HDC communications is the most apparent. A popular text on satellite communications describes how to build a cheap jamming installation by digging a round hole in the ground and covering it with wire. Imagine the outrage if a critic of United States foreign policy decided to jam an American communications satellite feeding the winning play sequence to millions of football fan homes during the Super Bowl!

Communications satellites are so important to deterrence and national defense that military and strategic planners recognize that they would be the first targets of an enemy attack. According to one published scenario, the satellites would be disabled or jammed, signaling the commencement of an actual full-scale nuclear strategic strike [49]. Paradoxically, the last region to become a battleground — outer space — would be the site of the first attacks in the ultimate war.

4. *Drawing the Lines in the Geostationary Battleground*

Outer space and, more specifically, the geostationary orbit, are becoming the primary areas of strategic conflict. The question remains whether the advantages offered by geostationary satellites to the superpowers and their alliance partners outweigh their vulnerabilities.

As stated earlier, communications satellites are a double-edged technological sword. On one side, they are a cost effective military force multiplier and a politically potent means for projecting national power. On the other side — the technology side — these same capabilities exact a technological and political cost. They invite strategic and tactical dependence on a rapidly evolving technology. This is a technology which uses two globally dispersed resources, the orbit and the spectrum resources, both existing outside the exclusive territorial sovereignty of any state. As such, it is a technology vulnerable to disruption or destruction by actors who can achieve their aims with a comparatively primitive and inexpensive technology and are outside a state's control. Politically, the military advantages enjoyed by the space powers through their use of communications satellites are accomplished *at the cost of becoming dependent on cooperative allocations and use of the orbit and spectrum resources* — cooperation of the world's smallest and weakest countries who command the majority of votes in the ITU.

The paradox is acknowledged in a United States Defense Department Directive that explicitly states the military policy towards satellite communications:

> Satellite communications are a vital transmission medium serving the Department of Defense. The capabilities which can be obtained and the services which can be provided by satellite communications are constrained by funds, available frequency spectrum, and optimal satellite orbital locations, among other constraints. [50]

The political cost at which the LDCs will allow the HDC military geostationary satellites to act as force multipliers in projecting national power, often against LDCs, is to a large degree determined by the rules promulgated by the ITU's World Administrative Radio Conferences and in other UN fora. The demands by the Group of 77 space "have-nots" at the 1982 United Nations Conference on the Exploration and Peaceful Uses of Outer Space (UNISPACE '82) to stop the continuing militarization of outer space by the superpowers illustrate growing dissatisfaction with the prospect of another global battleground literally hanging over each country's head.

In sum, the militarization of space is a strong politicizing influence on the 1985 and 1988 ITU Space WARC negotiations on access to the orbit spectrum resources. The majority of states that are technologically powerless to slow the expanding military use of the geostationary orbit will attempt to use Space WARC and their political voting power at the Conferences as a way of voicing their opposition to the militarization of space and satellite communications.

Economic Importance of Satellite Communications

Continued economic growth of the information societies is dependent upon the telecommunications networks made possible by communication satellites. For the United States and other technologically advanced countries:

> [A] large and increasing share of the gross national product of these countries are from information-related services as opposed to agricultural and manufacturing activities. Almost half of all United States economic activity is a result of the collection, organization, analysis, and dissemination of information and information-related services. Much of this is handled via microwave radio relay or domestic satellites. Thus the United States *has an ever-increasing dependence on the frequency spectrum and the geostationary orbit.* [emphasis added]. [51]

The telecommunications market itself spurs economic growth in other sectors. "The productivity of the United States telecommunications industry has grown more than twice as fast as productivity for the United States economy overall since 1950." [52] According to the Office of Technology Assessment (OTA) study, the United States has a 45 percent share of the estimated $250 billion world annual market for telecommunications, electronic, and computer equipment and services. This market is growing at an annual rate of 10-15 percent. Annual worldwide investment in telecommunications alone is projected to reach $88 billion by 1987. It is estimated that during the the world will invest over $640 billion in telecommunications equipment [53].

The growth rate of telecommunications equipment exports shows that a telecommunications "revolution" is taking place in many importing countries. Furthermore, a study of the global telecommunications market done by the consulting firm Arthur D. Little, Inc., projected that the most "dramatic" growth will take place in the satellite communications market sector. During the current decade, the study sees an additional 150-200 international earth stations (large, high capacity facilities) and thousands of smaller earth stations for regional and domestic systems of various capacities. In contrast to the general global recession, telecommunications exports are growing at about 7 percent per year [54].

1. The Information Societies

The information society is the electronic society. Technological breakthroughs that have fueled productivity prompt advanced economies to shift from a manufacturing based to a service-based economy. Two such breakthroughs were the invention of the transistor in 1947 and the development of the integrated circuit in the mid-1960s [55]. James Cook writes:

> Electronics is rapidly reshaping the world's industries in ways that turn tradition upon its head. In the process it is minimizing most of the elements that have always shaped industry in the past — labor, materials, manufacture. The basic thrust of the technology is immaterial and its productivity potential is enormous. After all, decision-making — the conversion of information into action — is probably the only really productive activity there is.[56]

Decision-making requires communications to put thought into action. As transportation, mailing, and printing costs have risen, telecommunications costs have declined. Teleconferencing is replacing business travel. American business spends $300 billion a year communicating information, and that will probably double by 1990. As the costs of conventional communications increase in price, the 1990 ($600 billion) expenditure will be predominately for electronic communications [57].

Since 1980, the Satellite Business Systems (SBS) — a consortium formed between International Business Machines (IBM), Aetna Life Insurance, and until 1984, COMSAT — has provided high capacity wideband telecommunications between widely scattered branches of large corporations. The following quote from an SBS advertisement illustrates the corporation's forward-looking futurist vision (which the company undoubtedly uses to justify their current operating losses):

> The story of SBS tells of a bold new corporate world of rooftop antenna dishes dotting city skylines and rural countrysides, of a new information age in which meeting rooms thousands of miles apart are linked through private network television, of the ability to send a volume of business data the size of Tolstoy's *War and Peace* anywhere in the country in one second.[58]

SBS claims three "firsts." It was the first commercial "all digital" satellite system using the less-congested Ku-Band (14/12 GHz) frequencies. Secondly, the system also employs advanced TDMA multiplexing (time-division multiple access — explained in Chapter 2). The absence of terrestrial frequency competitors, common to C-Band (6/4 GHz) users, allows placement of earth stations directly on the customer's premises, eliminating expensive ground links and other leased lines to an outlying earth station.

Additionally, SBS is capable of very high transmission rates, from 1.5 mb/s to 6.3 mb/s (mb/s–megabits per second; a bit is one piece of computer readable information). The advantage to a company that must transfer large volumes of data is illustrated by the following example:

> Suppose a company wants to send a 1-billion bit reel of data. Even if the sender uses what is generally the fastest speed available today, 56 kb/s, it will take 5.5 hours to send the date. The same file can be sent in 12 minutes at 1.5 mb/s.[59]

This capability permits companies to copy files quickly as backups against loss or erasure and to use under-utilized computers in different company divisions. Wideband communication is used for full video, voice, and data teleconferencing at "instant meetings," making much of today's expensive and time-consuming business travel obsolete. SBS estimates that large corporations spend 8 percent of their telecommunications budget for internal communications. If properly integrated into the company organizational structure, satellite communications promote greater efficiency of both human and material resources. In addition, it frees the company's organization and structure from geographical constraints. For these reasons, the major users of satellite communications are multinational corporations.

2. *Electronic Structuring of World Markets and Multinational Corporations*

The world marketplace is, in many respects, a global computer network. Multinational corporations (MNCs) were among the first organizations to realize that a global-scale "electronic office" would promote efficiency, allowing centralized administration of de-centralized manufacturing and marketing facilities. The worldwide telephone network being extended during the 1960s and 1970s to almost all the countries in the world through INTELSAT's intercontinental satellite circuits synergistically coincided with the MNCs' electronic administrative goals [60].

With the introduction of robotics to the assembly line, it is now feasible to directly monitor and control widely dispersed productive facilities through computer networks. The software for the robots can be stored in the administrative headquarters of the corporation. Politically, this allows the MNCs to invisibly control the transfer of technology while physically transferring the technology hardware to the buyer.

Technology now exerts a strong influence on the administrative structure of business organizations. The implications of this trend are seen in two directions; increased centralization of control through greater decentralization of control structures. In short, geographic distance has little meaning in the electronic office. Distance is measured in terms of time, and time is measured by the volume and speed of data streams between office computers.

There is a synergistic trend between technological innovations in microcomputers and telecommunications. As the costs for data processing and storage technologies decline, large "mainframe" computers are being challenged by the increasingly powerful "micro"-computers. Time-sharing systems where users "share" time on an expensive mainframe computer are no longer as economical given the cost and time efficiencies realized through use of the sophisticated microcomputers.

But time-sharing systems did have one advantage — they allowed users to share the same data base and to communicate findings and files among themselves. This advantage is rapidly disappearing as technological developments in private business exchanges (PBXs) convert previously analog telephone systems to all-digital networks in which voice, like data, is communicated in digital form. Today, each telephone jack in the electronic office is a computer connection linking desktop microcomputers together into local area networks (LANs) [61] sometimes characterized as integrated digital networks.

Users are both independent computers when doing individual work such as word processing; but they use their microcomputers also as "smart terminals" as part of the organization's mainframe computer when large-scale projects are undertaken. This marks the "wedding" of computers and telephones into the bonds of the so-called "value-added network."

Value-added networks utilize existing telephone circuits to interconnect users to data bases. TYMNET, TELENET, SITA, and SWIFT utilize INTELSAT circuits for their international links. [62] These networks "allow a faster, more reliable, and more cost effective link-up of computers than could be obtained if users were to lease standard lines." [63]

These developments have led to the electronic structuring of world markets. The evolution toward greater decentralization in the organization of multinational corporations dynamically coincides with the expansion of worldwide telecommunications networks. The United Nations Center on Transnational Corporations found that:

> The expansion of world trade and internationalization of such information-intensive industries as banking, insurance, tourism, and air transport intensified the need for mechanisms to ensure the instantaneous availability and dissemination of data ... The spread of transnational corporations, partly fueled, and partly dependent upon the creation of transnational computer communications systems that permit the rapid transmission of large volumes of data to control and coordinate large and complex organizations with functionally diverse and geographically dispersed operations. [64]

A worldwide multinational corporation uses the satellite-based global telephone network as its internal nervous system. Its ability to produce, sell, and trade goods and services among globally dispersed facilities requires real-time coordination and information. To quote a recent *Business Week* article:

> The single greatest impact on world communications is being made by digital computers. As these machines multiply [because of availability and lower costs] throughout business, telecommunications becomes an important strategic weapon to all companies. Instead of needing communications systems to transmit phone calls and telex messages, companies must now have them for such tasks as sending huge volumes of computer data at high speeds, transmitting facsimiles of blueprints, and holding video conferences. "Customers are demanding more of their communications systems, because, to be competitive in the world market, they have to automate more function," says Robert E. La Blanc, a New Jersey consultant and former vice-chairman of what is now Continental Telecom, Inc. ... Telecommunications has become the object of this mad scramble, because it has become so critical to the world's economy. The investment in communications plant and equipment by such industrialized countries as the US, France, and Japan already represents as much as

> 9 percent of their total gross domestic product, according to the ITU ... At the same time, a country's telecommunications infrastructure is becoming essential in maintaining a good business climate. Without modern, flexible, and competitively priced communications facilities, a country runs the risk of slowing overall economic development, if not stopping it all together. "Telecommunications are as essential to business infrastructure as good highways." [65]

However, with conventional commodities and services, there are protectionist and free-flow disputes regarding the transport of information goods and state economic sovereignty. The MNC is able, in many cases, to jump over state-imposed trade barriers by establishing production facilities behind quotas and tariffs erected against outside producers. As Third World countries become more aware of the key role transborder data flows (TBDF) play in information and technology transfer, as well as in the day-to-day management of MNCs, barriers are instituted to control data flows over national frontiers as a way of regaining prerogatives over MNCs. The issue of data flows has become salient to many policymakers in both industrial countries and MNCs:

> The politicization of the information issue is already apparent, and is a growing source of concern to United States policymakers: "Trade doesn't follow the flag anymore," according to a member of the United States Mission to the OECD, "it follows the communications system." [66]

The issue of free-flow of information and the concern of the Reagan Administration was echoed by Dr. Scott Thompson of the United States Information Agency:

> There should be an unrestricted flow of business information just as there should be unrestricted flow of goods and services in the world economy. There should be open and competitive information markets. [67]

In sum, MNCs depend on international computer systems to increase their maneuverability in relationships with government, suppliers, and customers. The MNC-computer interdependency is mutually reinforcing.

> Market incentives are, as always, guiding further technological developments in information processing. The large multinational firms with widely scattered operations represent the most attractive customers [for computer firms], so that enormous investments are being made to develop new applications and capabilities that would appeal to these firms. Further, the computer industry itself is becoming more unified worldwide through the use of common procedure, systems, and software. [68]

2.1. Control and Conflict over Transborder Data Flows

Telecommunications-based firms such as SBS, as well as many other MNCs, consider access to satellite communications and the orbit/spectrum resource of vital importance in keeping a competitive edge for the most efficient use of resources and coordinations of production. Future telecommunications needs are already intensifying conflicts between MNCs and host governments.

2.2. The Integrated Services Digital Network

The integrated services digital network (ISDN) could further reduce national governments' control over information flows between MNC operations. The ISDN will standardize all voice, data, telex, and even video telecommunications links to all-digital transmission modes. ISDN would interconnect computers worldwide in the same way the present telephone direct-dial system links countries and continents with standardized telephone dialing codes, billing, and technical transmission standards. ISDN would allow computers, telephones, data bases, and other information services throughout the world to communicate with each other directly — without translation from analog to digital. The information, though, appears the same — as binary bits or bursts of electromagnetic energy.

Digital convergence is bringing policy divergence as countries grapple with the problem of how to control this latest challenge to states' sovereignty over information flows. Like the militarization issue, data protectionism is one area where LDCs may seek bargaining leverage *vis-a-vis* the HDCs.

View from the North: A Political Assessment

Telecommunications and data flows have become "high" politics. As Edward Ploman, Rector of the United Nations University, writes in 1983:

> It is only in recent years that there has occured a shift in perspective so that communications and information by themselves have become issues in society, at the level of individuals and groups, and institutions, at the national and international level. All therefore face novel sets of issues. If proof were needed for how far this trend has gone it should be enough to recall the fall of governments mainly over broadcasting policy issues as has happened in Italy and the Netherlands or an international organization such as UNESCO almost being torn asunder over issues concerning the international flow of information. [69]

Within the political dimension, the issue of "prior consent" concerns HDCs and LDCs alike, politicizing the telecommunications and outer space policy environments. Since its inception as a technological possibility in the late

1960s, the direct broadcast satellite (DBS) has been the most viable point of contention between proponents of the "free flow of information" and those supporting "prior consent" for the regulation of transborder data flows of all types — DBS, remote sensing, *et cetera.*

Although since the earliest days of radio, international broadcasting has been both the practice and scourge of states, never before has it been technically possible to transmit high quality *television* on a global scale from a *single* satellite transmitter. In the absence of international legal, political, or technical prohibitions, any state can broadcast its television programs to viewers anywhere in the world.

Consequently, DBS instigates a lot of the continuing controversy over "prior consent." Some states contend they should retain the prerogative for "prior consent"; that is, before another state transmits programs from its satellite, consent must be obtained from the states receiving signal spillover. Prior consent is a new "right" of national sovereignty, created by satellite communications technology.

While international broadcasting has occurred both with and without recipient countrys' consent since the beginning of radio [70], DBS differs in two respects. First, the broadcasting is transmitted from an area outside of any state's territorial sovereignty, i.e., the geostationary orbital region of outer space.

Secondly, the DBS signal quality is better than conventional illegal "pirate" television and radio broadcasting. Once DBS systems are in place and operating, countries with state controlled and financed broadcasting worry that viewers will shift to foreign programming from satellites, or, with the choice of several satellites, will watch popular "foreign" programs financed with advertising revenues diverted from their own national media networks [71].

The issue of requiring prior consent for DBS broadcasts has served to further politicize the allocation and use of the geostationary orbit. Because DBS systems operate from a region legally recognized as part of the outer space commons, and outside of any state's sovereign jurisdiction, some states argue that they have a legal right to protest such broadcasts.

Use politicizes access. While "prior consent" deals specifically with program content, "technical consent" in the ITU fora refers to the technical parameters determining whether a signal will be receivable in a particular area. A technical parameter or standard can have the same effect as a control on the content, if the transmitted program, by adhering to the technical standard, cannot be received by an audience. Paragraph 6222 of the ITU Radio Regulations stipulates, in effect, a "technical consent" for DBS operations. Users of the geostationary orbit and radio frequency spectrum are to

> reduce, to the maximum extent practicable, the radiation over the territory of other countries unless an agreement has been previously reached with such countries. [72]

The question raised by Professor Carl Christol is central to the debate between use and access. He asks, "[D]oes a requirement for technical coordination constitute prior consent?" [73]

The debates over prior and technical consent reflect the political aspects of use permeating the normally technical negotiations concerning access to the orbit/spectrum resource [74]. The tight intermeshing of states' interests for both use and access produced the "plan" promulgated at the 1977 ITU World Administrative Radio Conference for the Satellite Broadcasting Service (designated as both WARC-77 and WARC-BS). At WARC-77,delegations concerned with the political ramifications of uncontrolled satellite broadcasting drew up detailed plans allocating orbital slots, channels, and service areas for DBS systems in ITU regions 1 and 3 (all areas of the world except North and South America). The countries in these regions favored this type of detailed orbit/spectrum resource planning because they saw the "plan" extending their control over what would otherwise be relatively unrestrained satellite broadcasting over their territories.

View from the South

The LDCs see satellite communications fulfilling basic and crucial telecommunications needs. Many developing countries consider satellite communications an especially appropriate technology for establishing national telecommunications infrastructures. There are numerous technical and financial advantages of satellite systems for the special developmental challenges of LDCs.

Developing countries face both natural and manmade barriers to the establishment of telecommunications systems. Deserts, mountain ranges, rivers, and extreme climatic conditions are forbidding natural phenomena that impede the construction and maintenance of conventional ground-based microwave relay systems. In these cases, satellite communications systems may be the most feasible solution.

Demographically, the distances between sparsely populated rural areas make the extension of urban telecommunications networks to these regions prohibitively expensive. In addition, the lack of trained maintenance personnel working with poor or nonexistent means of transportation render a terrestrial system vulnerable to disruption. When one remote link fails, it causes a major portion of the system to "crash." Reliability suffers due to the difficulty of mounting repair efforts to inaccessible relay points. Satellites bypass these natural and manmade problems.

The flexibility of satellite systems make them especially attractive to the LDC's telecommunications planners. A satellite system extends a telecommunications network wherever an earth station is placed; it is not limited to the particular location of the end terminal or route of a terrestrial cable or microwave relay system. Also, it offers a technological alternative to telecommunications planners and governments in their efforts to incorporate outlying rural regions into a cohesive national information infrastructure.

Politically, a satellite system may be the most expeditious method for establishing telecommunications in countries where unstable governments seek to reinforce their authority and legitimacy among diverse ethnic and social groups. The flexibility of satellite networks offer regimes the option of linking only those regions that support the political legitimacy of the government, while at the same time isolating regions of opposition. The LDCs are aware of the expanding potential of satellite communications for aspects of political, economic, and military power.

Political Aspects

It is interesting to note that the inauguration of global satellite services during the mid-1960s coincided with the period during which the countries in many former colonial regions gained their national independence from imperial powers. Most importantly, satellite communications enabled new countries to establish their telecommunications independence as well.

During and after the colonial era, it was commonplace in African colonies that even "local" telephone calls within the country were routed through a switching exchange in the former colonial capital, say London or Paris, located thousands of kilometers away [75].

For imperial powers, telecommunications cables were an important method to control colonies. These communications pathways persisted as the only reliable international and national telecommunications links until the advent of satellite networks. For former colonies, the large INTELSAT dish antenna became an important symbol of national independence and prestige, as they joined the global satellite telecommunications network. With satellite communications, countries were no longer dependent on former colonial powers for vital external and internal telecommunications circuits.

1. Political Legitimacy

A state is, in many ways,defined by its information flows [76]. A government's political legitimacy is affected by the amount, type, and source of information accessible to a country's inhabitants. An efficient and economical domestic telecommunications network is essential for political consolidation. Governments need communications to project power, to isolate areas of opposition, and to extend government-sponsored programs and services to maintain support. The education and socialization of each new generation requires the dissemination of information to all regions and ethnic and social groups.

The lack of telecommunications infrastructures in many of the LDC compounds the difficult task of integrating diverse groups of people into a politically cohesive state. Extending telecommunications to remote areas is a primary concern and developmental goal of LDCs.

1.1. A Star Over India

The Satellite Instructional Television Experiment (SITE) was a joint program between the United States National Aeronautics and Space Administration (NASA) and the Government of India. An Applications Technology Satellite (ATS-6) was moved in 1976 to an orbital slot over the Indian Ocean and its antenna pointed at the Indian subcontinent. The Indian government provided satellite television and radio receivers to 2,000 remote villages. These were operated with reliable electrical generators and maintenance-free antennas. Educational programs were transmitted via ATS-6 to areas previously without conventional telecommunications. "For the first time, all regions of the Indian subcontinent were electronically joined into a national information system. For the first time, these villagers saw live pictures and the voice of Prime Minister Indira Gandhi in their village. For the first time, there was a direct link between themselves and the centers of culture and political power." [77]

The SITE experience showed that satellite technology could electronically unite isolated and remote areas through a national information system. While the Indian experience was a temporary experiment to test satellite broadcasting and networking, Indonesia was the first Third World country to establish a permanent national satellite network in 1976.

In 1984, India placed its INSAT-1B into operation, providing national telephone, television, as well as weather reconaissance services. It was the first truly "hybrid" communications satellite to perform these functions from one orbital location. More importantly, it achieves a high degree of penetration into Indian society at a fraction of the cost of a conventional terrestrial system ("130 million dollars *versus* close to one trillion dollars")[78].

1.2. Indonesia's Palapa Satellite System

As the first LDC to install its own dedicated satellite system, Indonesia found communications satellites as the most workable means for linking the widely dispersed settlements, villages, and cities on the more than 1000 islands. Since 1976, a pair of Hughes HS-333 satellites, designed and constructed in the United States, have furnished the space segment telecommunications for the Palapa system. In the mid-1980s, Indonesia will augment its satellite capacity by procuring two additional Hughes HS-376 satellites to be launched on the American Space Shuttle. [79] Indonesia plans to use the new satellites to offer similar domestic telecommunications systems to other regional ASEAN states such as the Philippines.

At UNISPACE '82, Indonesia enjoyed international prestige as a pioneer among LDCs through its nationally owned and controlled satellite system for its own domestic needs. This prestige was enhanced because Palapa is operated independently of INTELSAT or any other outside operators.

Emulating the successful example of Indonesia, over 20 countries have announced their intention to establish dedicated national satellite systems for domestic telecommunications. Many of these countries are already leasing unused capacity on INTELSAT satellites for domestic services. But the fact that the least expensive INTELSAT transponders are available on a *pre-emptible* basis introduces an element of uncertainty, which however small, reinforces a perception of dependence or vulnerability of a country's sovereign prerogatives or control. For an aspect of sovereignty as central as control over the flow of information and telecommunications infrastructures, almost any degree of uncertainty imposes heavy dependency costs.

In sum, states perceive strong incentives for establishing *national* satellite systems. These include the choice of a system matching a country's particular needs, the lessening of uncertainty, vulnerability, and dependency on outside operators, and the prestige satellite technology represents. As a consequence, access to the orbit / spectrum resource as the requisite condition for the operation of a satellite system is an important foreign policy goal.

Satellite Communications and Economic Needs

Economic development requires communications. The LDC development strategists, faced with expanding populations and increasingly competitive markets abroad, are realizing the advantages to their country's economic development offered by satellite communications and information-processing systems. How do information and communications technologies aid economic growth and development?

Information is bargaining power. The economic competitiveness and bargaining positions of farmers and industrial firms alike improve if producers and sellers know world prices, capital goods and technologies, and the business and marketing practices of competitors.

Information infrastructures increase a state's economic bargaining, producing, and consuming power. Telecommunications puts information where it can more efficiently allocate scarce resources for production of goods and services. In countries with low literacy rates among rural populations, farmers can learn to apply advanced planting techniques to increase crop yields through television programs beamed directly at a village receiver from a satellite in the geostationary orbit. In the same fashion, distributors of agricultural products can learn the most current national and world prices for the crops they will bring to markets. Educational programs that can help fight disease, control flooding, and teach farm children how to read can all be disseminated to literally thousands of remote villages from a single satellite. For these reasons, the LDCs envision satellite systems as technological "leapfrogs" to space age economic development; a technology that allows these countries to skip over obsolete information and communications technologies [80].

1. Impact of Satellite Communications on Development

While the technical and financial advantages of satellite communications systems are apparent, there are as yet few studies analyzing how telecommunications further economic development. An ITU/OECD study released in 1982 found that the relationship between benefits and costs of investments in telecommunications and economic growth approached 200:1 for manufacturing industries in Kenya and 85:1 in rural Egypt [81]. The specific effect of satellite communications was estimated by Dr. Heather Hudson, *et al.*, to bring about an approximate $370,000 increase in Gross Domestic Product (GDP) for each earth station providing telephone service to previously unconnected rural regions [82]. Other studies have found:

> . . .that a 1.0 percent rise in the number of telephones per 100 population between 1950 and 1955 (in 52 countries studied) contributed to a rise in per capita income, between 1955 and 1962, of about 3.0 percent. The poorer the country, what is more, the greater the benefit, and — against all expectations — the home telephone was a bigger economic boon than the telephone in a place of business because the home telephone is available 24 hours a day. There are also social benefits. In rural areas the telephone is used for more for emergencies than it is in towns. [83]

According to the Hudson study, telecommunications is a major promoter of growth and development for the following reasons:

> [It] improved management efficiency and productivity through such applications as inventory control, fleet scheduling, and economies of scale. [Telecommunications are used] by rural producers who need to know current prices offered by various buyers and to locate markets for their products. [Telecommunications can] reduce spoilage of perishable products [and] to order supplies and parts so that down time of rural equipment [is reduced].

> Telecommunications can also play an important administrative and coordinating role in supporting rural projects. Regular communication between headquarters and the field is vital for the management of any rural project. [84]

1.1. Satellite Telecommunications for Rural Areas

The major problem that stands in the way of LDC satellite networks is, in short, money. The Maitland Commission noted:

> 41. If developing countries are to give greater priority to telecommunications and to improve and expand their public telecommunications networks, we estimate that total investment of US $12 billion a year will be needed. [85]

Telecommunications, while shown to be a factor promoting development, costs a lot, especially when extended to rural areas — the areas most in need of telecommunications services the problem and promise is clearly posed by an ITU/OELD study which found:

> In the developing countries, communications in the rural areas are one of the most pressing problems today. In these countries typically 80 % of the population is rural. At the same time, frequently these rural areas constitute the major source of foreign currency earnings; the exports being produce of various kinds, minerals, processed foods, etc. [86]

1.1.1. Telecommunications for Rural Areas and Economic Development

Rural telecommunications to sparsely populated, outlying agricultural regions cost more to establish and support than urban systems. [87] ITU Secretary-General Butler points out:

> The traditional method of telecommunication transmission has a cost factor that increases with distance. It also requires a maintenance effort that increases with the length of the system, *but this is not so for satellite systems* [emphasis added]. Given the high costs coupled with low traffic density, telecommunications installed in rural, isolated and underpriviledged areas were not *immediately profitable* [sic]. Thus, in the absence of the historical rate of return for the service, central planners, banks, and other financial institutions which look to a systematic return on loans or capital have been unable to give a high priority for investment capital. Indeed, it is users who have sustained a large part of the investment. [87]

Cost problems are exacerbated by the income level of rural residents and communities. The major financial barrier to extending telecommunications to outlying regions lies in the difficulty of users in rural areas to pay higher rates where costs are sensitive to distance and subscriber density. Predictably, the result is an unequal disribution of telecommunications services between urban and rural areas.

In the three LDCs listed below, per capita gross national product comes to less than $300 (US) per year, or about $8 per month. The resulting maldistribution of telecommunications services is depicted in Table 1.1, which shows demographic and telephone distribution.

Table 1.1
Demographic Distribution of Telephones in Three LDCs

Country	% of Population	No. of Telephones	Telephone Density per 100 Population
Ethiopia	100.0	79,000	0.27
Addis Ababa	4.5	46,524	3.50
Other Urban	1.5	11,362	2.64
Rural Areas	94.0	21,114	0.08
Haiti	100.0	17,800	0.37
Port-au-Prince	7.9	10,625	2.90
Other Urban	16.2	830	0.11
Rural Areas	75.9	6,345	0.18
Tanzania	100.0	66,300	0.43
Dar Es Salaam	2.1	29,326	9.24
Other Urban	0.9	6,018	4.60
Rural Areas	97.1	30,956	0.21

Source: International Telecommunication Union, Report of the Administrative Council, Annex: Review of the State of Telecommunication Services in the Least Developed Countries and Concrete Measures for Telecommunication Development. Document No. 48-E, Geneva, May 1982, p. 23.

1.1.2. Funding of Rural Telecommunications

Telecommunications are often a low priority goal in LDC development programs — for financial and political reasons. LDCs with low per capita GNP cannot generate domestic savings or enough capital for large-scale investments in telecommunications systems. In scarce capital markets, highly visible transportation, health, and security services compete against telecommunications services for investment funds. Telecommunications, being less recognizable in finished products and services and perceived as a luxury item for wealthy urban consumers, often holds a lower priority for governmental development funds.

The financial dilemma is compounded by the fact that advanced telecommunications equipment must be purchased abroad. Limited foreign currency reserves (earned from imports) severely limit the amount and link of telecommunications equipment that can be purchased [88]. High costs and the perception that the telephone and television are luxuries compel many of the LDC governments to use telecommunications primarily as a source of revenue for the government-owned telecommunications monopoly.

2. Economies of Satellite Communications for LDCs

Although initially "expensive," satellite networks offer many cost advantages over conventional terrestrial networks. For these reasons, satellite networks are especially attractive to telecommunications planners in the many Third World countries with thin or widely dispersed populations. While satellites may not currently offer the most practical technology for all developing areas or countries, the rapidly declining cost of satellite hardware will soon present alternatives that are financially feasible for many of today's "marginal" countries.

To briefly review the economic advantages of satellite-based systems, they are as follows:

1. The cost of satellite telecommunications is distance-insensitive. One satellite transponder can broadcast a high-quality television or radio program to an entire country, linking simultaneously thousands of villages and cities.
2. Maintenance costs do not increase with the length or distance apart of system nodes.
3. The satellite offers a high degree of planning flexibility. The earth stations providing telephone and other services can be located where the people live and work, not where geography dictates or where the cable link ends.

The insensitivity of satellite costs to distance makes satellite networks perhaps the only practicable means for establishing rural telecommunications infrastructures in many LDCs. Significantly, the economies of satellite communications closely match the demographic and development strategies of many large rural Third World countries. Many of the LDCs express their goals of telecommunications development strategies in telephones per kilometer, rather than by telephones per capita [89].

2.1. GLODOM: Satellite Telephones to Rural Areas

Recognizing the value and ability of satellites to connect isolated geographical and urban-rural regions, the Global Domestic Satellite System Program (GLODOM), as studied by the ITU and promoted by its Secretary-General Richard E. Butler of Australia, actively seeks to extend satellite solutions for regional and national telecommunications needs [90].

GLODOM sees the use of common-user communications satellites providing domestic telecommunications in countries occupying a particular geographical region. Under the GLODOM arrangements, countries within a region would share the capacity of a single satellite, each utilizing different transponders for their own particular needs and specific system technical parameters.

Inexpensive earth stations form the core of the GLODOM concept. In this way, GLODOM follows the evolutionary cost inversion that has taken place since Early Bird, where higher transmitter power levels and increasing investment in space has resulted in reduced costs and earth station complexity on the ground. Today, satellites can transmit programs or data with high power transponders so that small, inexpensive antennas would suffice for acceptable signal quality [91]. Portable, inexpensive, easy to maintain, and with a highly reliable self-contained power supply (Butler has proposed solar panel batteries), the GLODOM earth station would effectively free telecommunications planners to extend the system anywhere as needed.

2.2. Rural Satellite Telecommunications and the Orbit/Spectrum Resource

The cost of the earth station is the most crucial factor for the establishment of a rural satellite network. As we have seen, space segment costs do not increase with distance between communicators making communication satellites an economical means for long-distance, low-volume telecommunications links to outlying regions. Since satellite hardware costs are fixed by technical requirements imposed by the harsh operating conditions of outer space and by the ITU technical performance and interference standards, telecommunications planners are compelled to economize in the design and operating characteristics of the ground segment. Even relatively small savings in the cost of an earth station are magnified in cost calculations involving hundreds or thousands of earth stations.

The crucial economic considerations for earth station planning are the technical requirements and interference parameters which determine how efficiently the earth station must use and conserve the orbit/spectrum resource. Conservation of the orbit/spectrum resource requires more stringent performance standards and more expensive technology. These costs rise as the orbit/spectrum resource become more crowded, compelling the utilization of resource-conserving technologies and earth station design.

Most, if not all, LDC satellite systems use or will be using the most crowded frequency band — the C-Band (see Chapter 2 for the exact frequencies). For many countries in the tropical latitudes, the C-Band is currently the most desirable frequency band due to its imperviousness to rainfall. The signal degradation in the higher frequencies during heavy rainfall makes the less crowded K-Band frequencies unfavorable for tropical countries.

We see, therefore, a trade-off between orbital spacing of satellites with overlapping beam footprints and the cost of earth station antennas. Small, inexpensive earth station antennas require wider spacing between satellites in the geostationary orbit to prevent harmful interference. In view of the LDCs' wish to establish rural satellite networks, to keep earth station costs low, orbital spacing between satellites must be apportioned in a manner that will accommodate small inexpensive antennas [92].

3. LDC Needs and Demand for the Orbit/Spectrum Resource

The LDCs are today, and will be for the future, the source of the greatest demand for satellite communications and portions of the orbit/spectrum resource as they attempt to meet their present and projected telecommunications needs. Dr. Joseph Pelton, Executive Assistant to the Director-General of INTELSAT, emphasized the need in the Third World for telecommunications and implications for demand of the orbit/spectrum resource by stating,

> In order to bring telecommunications to isolated rural populations, there will have to occur a "Communications Revolution". [93]

The questions are explicit: what acceptable level of telecommunications service distribution, at what cost, and with which means — microwave link, cables, or satellites — will the telecommunications needs of the LDC populations be met? What are the LDC needs? Do they require a level of telecommunications service commensurate with that found in the HDCs? The Maitland Commission report, taking into account all the financial, political, and technical constraints along with the current level of services in LDCs, came up with the following goal:

> We believe that by the early part of the next century virtually the whole of mankind should be brought within easy reach of a telephone and, in due course, the other services telecommunications can provide. That should be the overriding objective. [94]

The poor of the world have a long way to the next phone. The table below shows the maldistribution of telecommunications services and facilities between the HDCs and the LDCs:

Table 1.2
Telephone Density

Country (HDCs)	Telephone Density No./100 Population
United States	77.0
Sweden	74.4
Switzerland	68.2
Canada	63.5
Denmark	56.5
Japan	45.8
West Germany	40.3
East Germany	17.7
Poland	8.9
Country (LDCs)	**Telephone Density No./100 Population**
Afghanistan	0.17
Ethiopia	0.27
Haiti	0.37
Sudan	0.33
Tanzania	0.43

Source: International Telecommunication Union, Report of the Administrative Council to the Plenipotentiary Conference, Review of the State of Telecommunication Services in the Least Developed Countries and Concrete Measures for Telecommunication Development, May 1982.

The leaders of many LDCs, in international fora, loudly proclaim the advantages of satellite communications for attaining political and economic goals. As the above table shows, bringing the LDCs telephone service up to the level enjoyed by the HDCs would require satellite capacities far beyond what is technically possible today.

The Gap Between North and South

The global direct-dial telephone network is the largest and most complex machine ever constructed. It is still in a relatively primitive and uneven stage of development; certain sections are more sophisticated due to their greater traffic capacities and penetration into societal and geographic groupings and regions. In the Third World, the financial costs of extending networks and services to bring coverage and capacity up to present HDC standards would literally be astronomical. A telecommunications financial analyst has estimated that investments in telecommunications must be at least $50 billion a year for 30 years in order to bring the rest of the world up to *current* United States standards [95].

In the meantime, the United States and other HDCs are experiencing explosive growth and development in their own telecommunications markets and information processing technologies. In the United States, where there are already 95 telephones for every 100 households, 6.5 million a year are still added to its system, an increment that is greater than all the telephones in all of Africa (outside South Africa) [96]. Currently, there are about 500 million telephones in use worldwide. According to experts, the minimum acceptable global system would consist of about one billion telephones, which would cost about $750 billion by the 1990s or one trillion dollars by the year 2000. Buddy, can you spare some change? There's a call I wanna make...

Summary

This chapter has reviewed how and to what purposes countries use satellite communications and why they desire access to the orbit/spectrum resource for their own satellite systems. While it is difficult to assess the specific power value satellite systems represent, it is nonetheless apparent that communications satellites do operate synergistically with military forces, multinational corporations, global banking networks, and other information technologies as "projectors of power." For these reasons, many LDCs perceive the asymmetrical distribution of information technologies (including satellites) between themselves and the HDCs as widening the power gap between North and South.

That communications satellites are perceived as widening the gap is somewhat of a paradox, for it is satellite systems which brought modern telecommunications to many Third World countries. But a growing awareness of the economic, cultural, political, and strategic power roles that information technologies play, combined with a growing discrepancy in the development and use of these networks between North and South, mean that communications satellites are increasingly perceived as working to widen the global power gap. There is much evidence showing quite clearly the extent to which the information technology gap between North and South *is* growing; while the HDC telecommunications planners worry about meeting demand with high capacity fiber optic networks, their LDC counterparts are concerned with how to get *one* telephone into remote villages. Even though some LDCs are making progress in this area, the vast majority are falling further and further behind.

In short, the flexibilities and economies of communications satellites make them a most feasible and applicable information technology to meet many of the LDC needs. These needs, overlaying perceptions of a tightening bond between information competence and national power, have focused many states' attention on satellite systems, and concomitantly, on access to the orbit/spectrum resource. The present importance of satellite systems magnifies projections of their future economic, political, and military power role. This has resulted in a great increase in the number of countries planning their own satellite systems, a proliferation that, barring revolutionary breakthroughs in technology, may greatly increase crowding of the most favorable portions of the orbit/spectrum resource.

Perceptions of Resource Scarcity

The implications for orbit/spectrum use are clear. The GLODOM Proposal envisages 48,000 earth stations in LDC remote villages and small towns by 1990. In order to accommodate, or to minimally satisfy the future demand for telecommunications in the Third World, either new sections of the resources must be opened by technological advance, or more efficient means must be applied to currently used portions of the orbit/spectrum resource. In any case, the rising demand and need for telecommunications serves to intensify LDC perceptions of orbit/spectrum scarcity. Politics is often the power behind symbols and the prestige associated with being part of the Space Age. Access to global information flows has, in turn, politicized access to the orbit/spectrum resource.

REFERENCES

1. William G. McGowan, Chairman, MCI Communications Corporation, and J.S. Malecela, Minister for Communications and Transport, Tanzania, both quoted in Special Advertising Supplement, *Time*, October 1983.
2. INTELSAT, *INTELSAT Handbook*, 1983, sec. 4.
3. Jeffrey M. Lenorovitz, "Soviets Marketing Proton in West," *Aviation Week and Space Technology*, June 20, 1983, pp. 18-20; and, "Soviets Provide Data to Guide INMARSAT in Launcher Decision, *Aviation Week and Space Technology*, August 8, 1983, p. 22.
4. This was the prognosis made by John Kiesling, President on the Mobile Satellite Corporation at the International Business in Space conference, January 9, 1985, Washington, DC. See Stephen Doyle, "INMARSAT: The International Maritime Satellite Organization — Origins and Structure," *Journal of Space Law*, 5:1(1977):45-64; and H.H.M. Sondaal, "The Current Situation in the Field of Maritime Communication Satellites: 'INMARSAT'," *Journal of Space Law*, 8:1(1980):9-39; and "INMARSAT Plans Aircraft Communication", *Aviation Week and Space Technology*, November 15, 1982, pp. 79-84.
5. See National Paper: Soviet Union, presented at UNISPACE '82, August 1982, Vienna. A/CONF.101/NP/30; and Theo Pirard, "INTERSPUTNIK: The Eastern "Brother" of Intelsat," *Satellite Communications*, August 1982, pp. 38-44.

6. Joseph Pelton, *et al.*, "INTELSAT: The Global Telecommunications Network", paper presented at the Pacific Telecommunications Conference, Honolulu, January 1983, p. 5.
7. INTELSAT, "Domestic Leases: Cost-Effective Flexible National Satellite Networks," booklet distributed at the Pacific Telecommunications Conference, Honolulu, January 13-17, 1985.
8. Richard Colino, "INTELSAT and the Future of the Orbital Arc," paper and speech presented at the Pacific Telecommunications Conference, Honolulu, January 15, 1985.
9. "INTELSAT macht Telefonieren billiger," *Astronautik*, 2(1981), p. 33.
10. Comments by David Leive during his presentation at Telecom '83, Geneva, October 30, 1983; author's notes.
11. Sources: INTELSAT, *INTELSAT Handbook*, 1983, Appendix 6; Interview with Dr. Henry Meyerhoff, IFRB, Geneva, May 27, 1982.
12. "Telecommunications Expert Criticizes ITU, INTELSAT as Club Dominated by Big Firms," *Communications International*, April 29, 1983; interview with Professor Lusignan, Stanford University, Palo Alto, California, February 7, 1984; author's notes.
13. Interview with Dr. Bruce Lusignan, 1984, Stanford University, Palo Alto, California, February 7, 1984.
14. See, Eilene Galloway, "The United States and International Space Cooperation," unpublished paper prepared for the Office of Technology Assessment, November 9, 1982, p. 39.
15. INTELSAT, *INTELSAT Handbook* (1983), Appendix 6.
16. Jay C. Lowndes, "INTELSAT Alters Earth Station Standards," *Aviation Week and Space Technology*, January 16, 1984, p. 203.
17. See, *Background Paper: Arab Satellite Communications Organization*, A/CONF.101/BP/EGO/4, July 8,1981, distributed at UNISPACE '82, August 1982, Vienna.
18. Source: Ali Al-Mashat, "The Arab Satellite Communications System," *AIAA 9th Satellite Communications Conference Proceedings*, San Diego, California, March 1982, pp. 187-191.
19. Source: EUTELSAT Brochure distributed at Telecom '83, Geneva, Switzerland, October 1983.
20. In reality, ARABSAT is a region's realization of the benefits accrued by Indonesia as the first LDC to own a national satellite network for its domestic telecommunications. The "Palapa" system demonstrated the technical, economic, and political advantages from having a dedicated satellite system designed with a particular country's needs in mind.
21. Ali Al-Mashat, 1982, p.191.
22. S. Brown, *et al., Regimes for the Ocean. Outer Space, and Weather,* Washington: Brookings Institute, 1977, p. 176.

23. See, UNESCO, *Many Voices, One World: Communication and Society Today and Tomorrow*, Paris, 1980; and, Kusum Singh and Bertram Gross, "The MacBride Report: The Results and Response," in George Gerbner and Marsha Siefert, editors, *World Communications: A Handbook*, New York: Longman, 1984, pp. 445-456. The MacBride Commission studied the patterns and effects of global news reporting, mass media, and other information flows — particularly between HDCs and LDCs — on Third World economic, political and cultural development. The report calls for a "New World Information and Communication Order," a phrase coined by the Tunisian delegate, Mustapha Masmoudi. See Mustapha Masmoudi, "The New World Information Order," in Gerbner and Siefert, editors, pp. 14-27, reprinted from the MacBride Commission Report cited above.
24. The underdevelopment of the Soviet bloc information competence was graphically illustrated to this author on an October 1983 tour of an Hungarian Geodetic Observatory, located about 35 kilometers outside Budapest. Upon seeing the primitive computing facilities (one researcher was using a Timex/Sinclair ZX-81 computer) a fellow visitor who was on the editorial board of *Spaceflight*, the British Interplanetary Society Magazine commented, "Well, to say what we really think would be to kick them in the teeth, for they are trying their hardest." Tour taken on October 11, 1983 during the International Astronautical Federation Congress in Budapest, Hungary.
25. Anthony Smith, *The Geopolitics of Information: How Western Culture Dominates the World*. New York: Oxford University Press, 1980, pp. 111-147.
26. Ronald Stowe, "The Legal and Political Considerations of the 1985 WARC," *Journal of Space Law* 11:1-2(1983):63.
27. Edward Miles, "Transnationalism in Space: Inner and Outer," *International Organization* 25(1971):604.
28. Dallas Smythe, *Dependency Road: Communications, Consciousness, and Capitalism in Canada*. Norwood: Ablex, 1981, p. 317.
29. Dennis McDougal, "U.S. Broadcasts to Cuba Miss Targeted Start-Up," *Los Angeles Times* January 30, 1985. See Robert A. Kinn, "United States Participation in the International Telecommunication Union: A Series of Interviews," The Fletcher Forum, Winter 1985, p. 65.
30. Author's notes from Telecom '83, Geneva, October 30, 1983.
31. Craig Covault, "Space Command, NORAD Merging Missile, Air and Space Warning Roles," *Aviation Week and Space Technology*, February 11, 1985, pp. 60-62; and Ashton B. Carter, "The Command and Control of Nuclear War," *Scientific American*, January 1985, pp. 32-39.

32. Richard Halloran, "United States Turning Its Attention to the New Theater of Military Operations: Space," in *International Herald Tribune*, October 20, 1982.
    A representative sampling of non-classified US military satellites in geostationary orbit includes the following:
    "Vela," twelve satellites that can detect nuclear explosions with infrared sensors.
    "DSCS II," fourteen satellites and two spares operate with high power and wide (410 MHz) bandwidth.
    "FLTSATCOM," five satellites for high-priority UHF communications for land terminals, Strategic Air Command, and the Presidential networks.
    "AFSATCOM," twelve channel communications packages on the FLTSATCOM satellites for Air Force traffic.
    "DSCS III," a fleet of twelve are planned to replace the aging DSCS II satellites for X and UHF band communications.
    "Milstar," a multi-billion dollar project now under development by Lockheed Corporation, is envisioned as a system composed of approximately four satellites in geostationary orbit and three in polar orbits. A project manager for Hughes Aircraft stated in an off-the-record interview in October 1982 that the satellites are hardened to withstand and survive the effects of electromagnetic pulse "when there's no one left to communicate with." The Milstar satellites are designed to operate autonomously without continuous ground-based control, a capability necessary during a nuclear war where the survivability of ground-based control facilities would be in doubt. See, James Fawcette, "Milstar: Hotline in the Sky," *High Technology*, November 1983, pp. 62-69; and, Tom Canby, "Satellites That Serve Us," *National Geographic*, September 1983, p. 331.
33. For example, the Soviet Union is reportedly using satellite links to connect radar experts in the Soviet Union with SAM-5 anti-aircraft missile emplacements in Syria. The Soviet experts can, from within the safety and political isolation of their own country, direct missile defenses located thousands of kilometers away. See *Los Angeles Times*, April 10, 1983.
34. James B. Schultz, "Reliable Survivable Satellites Seen as Key Link in United States Security," in *Satellite Coomunications*, June, 1980, p. 26.
35. Schultz, p. 30.
36. Richard C. Henry, Book review of Thomas Karas, *The New High Ground — Strategies & Weapons of Space Age War*. New York: Simon and Shuster, 1983; review appeared in IEEE *Spectrum*, September 1983, p. 9; see also, Tom Karas, "Go-codes from Out There: Space Communications for World War III", *Technology Illustrated*, March 1983, pp.32-40.

37. Brochure published by Rockwell International Corporation, 1983.
38. Michael May, "War or Peace in Space," monograph published by the California Seminar on International Security and Foreign Policy #93, March 1981, p. 2.
39. See, Robert B. Giffen, "US Space System Survivability," *National Security Affairs Monograph*, Series 82-4, National Defense University Press, Washington, DC, 1982, (hereafter cited as Giffen).
40. This discussion is guided by Giffen's outline of satellite security, pp. 26-29.
41. Comments by Thomas Karas on Public Broadcasting Service program entitled, "The Race for the High Ground," broadcast on "Frontline," April 13, 1983.
42. Giffen, p. 10; see also, John Mattox, "The Strategic Implications of the Electromagnetic Pulse Phenomena," paper submitted to the 1983 Student Pugwash Conference on Science, Technology, and Global Responsibility, University of Michagan, Ann Arbor, June 1983; J.J. Hutter, "EMP and International Security," paper presented to the International Student Pugwash Conference on Science, Technology, and Global Responsibility, University of Michigan, Ann Arbor, June 1983; and William J. Broad, "A Fatal Flaw in the Concept of Space War," *Science*, March 1982, pp. 1372-1374.
43. Elektroschocks aus dem All," *Die Zeit*, January 15, 1982.
44. "Bomb to Cripple Communications Studied." *Los Angeles Times*, April 17, 1983, Part I, p. 14.
45. Giffen, p. 28. Dr. Robert Bowman, in an informal interview on October 10, 1983, in Budapest, Hungary, stated that identification of each piece of the increasing amount of "space junk" — rocket motors, protective casings, exploded parts, and shreuse.
46. Interview with Professor Bruce Lusignan, Stanford University Communications Satellite Planning Center, February 7, 1984. See also, "Arabsat Satellite's Control Signals Will Be Encrypted," *Aviation Week and Space Technology*, May 21, 1984, pp. 176-177.
47. Giffen, p. 29.
48. Giffen, p. 30. The vulnerability of the United States ground segment is illustrated by the location of large Air Force Satellite Command dish antennas situated within 100 meters of a freeway in Sunnyvale, California. This proximity makes them not only a target for radio wave jamming, but also to direct physical attack on the antenna structures themselves. Personal observation by author in March 1983.
49. Giffen, p. 1.
50. US Defense Department Directive: Military Satellite Communications Systems Organization (MILSATCOM), No. 5105.44, October 9, 1973.

51. *United States Congress, Office of Technology Assessment*, "Radiofrequency Use and Management: Impacts from the World Administrative Radio Conference of 1979," January 1982, p. 27.
52. *Ibid.*, p. 28.
53. *Ibid.*
54. Statement by Edgar A. Grabhorn of Arthur D. Little, Inc. Director of World Telecommunications Information Program, in *Inteltrade*, October 15-30, 1982, p. 1.
55. ENIAC, the first large computer, had 18,000 overheating vacuum tubes and cost $3 million in the late 1940s. The comparable machine today costs less than $100 and can fit in a coat pocket.
56. James Cook, "The Molting of America," *Forbes*, November 22, 1982, p. 163.
57. "Toward the Wired Society," *World Business Weekly*, June 8, 1981, p. 31.
58. Quoted from advertisement in *COMSAT Magazine*, December 1980.
59. *Ibid.*, p. 4.
60. "Electronic Mail," *Los Angeles Times*, February 24, 1985. The article points out how dependent a large decentralized multinational corporation is on electronic mail, with the example, "how would the administrative staff be able to contact the 8,000 franchises of a fast-food chain so as to warn them of worms in the hamburger meat," other than through electronic means.
61. Although a LAN may be "local" in a geographic sense, there is no technological constraint for any distance limitation. In essence, the globe is now considered local; "local" now refers to the degree of software interaction with the various CPUs.
62. TYMNET and TELENET are value-added networks that allow different types of computers to exchange and relay information and data bases among their network nodes at high rates of speed and with a large degree of flexibility. SITA (Societe Internationale de Telecommunications Aeronautiques) and SWIFT (Society for Worldwide Interbank Financial Telecommunications), as computer networks for airlines and banks, respectively, also rely on dependable high-quality data links.
63. "Transnational Corporations Dominate Transborder Data Flows," in *Intermedia*, May 1982, p. 49. See also, Gareth Locksley, "The Political Economy of Satellite Business," *Telecommunications Policy*, September 1983, pp. 195-203.
64. *Ibid.*, p. 48.
65. "Telecommunications: The Global Battle," *Business Week*, October 24, 1983, pp. 62-65.

66. Rita Cruise O'Brien and G.K. Helleiner, "The Political Economy of Information in a Changing International Economic Order," *International Organization*, 34:4, Autumn 1980, p. 457.
67. Quoted from remarks made by Scott Thompson at Pacific Telecommunications Conference, Honolulu, January 16, 1983.
68. Cruise O'Brien, p. 466.
69. Edward Ploman, "National Needs in an International Communications Setting," paper presented at Pacific Telecommunications Conference, Honolulu, January 16, 1983. Ploman correctly assessed the situation, as the United States withdrew from UNESCO in 1984, and the United Kingdom announced its intention to do the same in 1985.
70. See, Delbert D. Smith, *International Broadcasting*. New York: Sijthoff Publishing, 1969.
71. This pertains particularly to the Federal Republic of Germany's (FRG) policies regarding Radio Luxembourg's (RTL) plans to Operate a DBS with a beam footprint covering most of the FRG. To put it simply, satellite broadcasting places West German information policy in a quandary. On one hand, the FRG favors the free flow of information since its DBS will illuminate the German Democratic Republic (GDR). On the other hand, it is feared that the RTL American-style commercial mass-appeal programming will draw West German viewers and advertising revenues away from the West German networks and print media. In anticipation of this, West German publishers have provided 25 percent of the RTL satellite project financing. See, Institut Fur Rundfunkrecht, Universitaet Koln, 1982 Symposium on Direct Broadcast Satellites; and, Marcellus Snow, "West German Media Politics," *Journal of Communications*, Summer 1982.
72. ITU, *Final Acts*, 1979 WARC. Richard E. Butler, Secretary-General of the ITU, remarked that as far as the ITU was concerned there was "no problem with the prior consent issue," since the ITU recognized only a technical consent. Interview on December 13, 1982 in Geneva, Switzerland.
73. Carl Q. Christol, "Telecommunications, Outer Space, and the NIIO," *Syracuse Journal of International Law and Commerce* 8:2(1981):352.
74. This was the topic of an exchange between Costa Rican telecommunications lawyer Lilianna Garcia de Davis and other panelists in response to this researcher's question regarding the linkage between "use" and "access" negotiations. Garcia de Davis stated that in the LDC view, free flow translates to inundation of their networks by the HDC programming sources. Forum 83, October 29, 1983, Geneva.

75. The INTELSAT system has had the effect of decentralizing telecommunications and the importance of former switching centers. As Wilson Dizard points out,
"[T]he satellite network has reduced the importance of the "gateway cities" in Europe, the United States, and Japan through which communications traffic to the rest of the world was formerly channeled. Until the advent of the satellite network, for instance, all international traffic between the east and west coasts of South America was routed through gateway cities in the US. Similar changes in Asia and Africa have given Third World countries direct communications with each other for the first time, thus lessening their dependence on the industrialized nations." Wilson Dizard, *The Coming Information Age: An Overview of Technology, Economics, and Politics*. New York: Longman, 1982, p. 150.
76. Karl Deutsch, *The Nerves of Government*. New York: The Free Press, 1963.
77. Interview with Richard Nickelson, Senior Counselor, CCIR, ITU, Geneva, May 29, 1982. Nickelson, the program director, stated that in addition to the educational purposes, SITE offered important political advantages to the Gandhi government. The government-run national television channels now had access to many otherwise hard-to-reach areas and groups. The government in power had the opportunity to associate itself with the prestige and wonder the people held for the technological "miracle." On the other hand, other observers reported that many villagers left after the soap operas, having evidently more important things to do than watch the Prime Minister give a speech. The political ramifications were well documented in the Public Broadcasting Service "NOVA" report on the "Global Village: Satellite TV in India," which aired in January 1985.
78. PBS, "The Global Village..."
79. On February 6, 1984, one of the newer Hughes HS-376 satellites purchased by Indonesia failed to reach the geostationary orbit when its McDonnell-Douglas payload assist module (PAM) misfired. Although insured, the loss will be a setback to the Indonesian plans to extend their telecommunications capacity to other countries. In the fall of 1984, a daring Space Shuttle salvage mission successfully recovered the two errant satellites (a similar Hughes satellite owned by Western Union also had misfired).
80. See, ITU, "The Missing Link: Report of the Independent Commission for World Wide Telecommunications Development," 1984.
81. See, W. Pierce and N. Jequier, "Telecommunications and Development: General Synthesis Report on the Contribution of Telecommunications to Economic and Social Development," ITU/OECD, Geneva

and Paris 1982; Heather E. Hudson, Andrew P. Hardy, Edwin B. Parker, "Impact of Telephone and Satellite Earth Stations Installations on GDP," *Telecommunications Policy* 6:4, December 1982, pp. 300-307.

82. Hudson, *et al.*, p. 304.
83. "Third World Telephones: Brother Can You Spare a Dime?", *The Economist*, December 17, 1983, quoted from *Teleclippings*, January 16, 1984, pp. 2-3.
84. Hudson, *et al.*, p. 306.
85. ITU, "The Missing Link:. . ." (1985) Executive Summary, p. 13.
86. Richard E. Butler, "Satellite Communications for Developing Countries," monograph presented at the United Nations Interregional Seminar on Space Applications in Preparation for UNISPACE 82, Addis Ababa, 18 June 1982, p. 2.
87. *Ibid*, p. 4
88. ITU, *Report of the Administrative Council*, p. 43.
89. In the ITU Administrative Council Report, telecommunications analysts defined "minimum needs" as one telephone for a population cluster within a 5 kilometer radius. "In most of the countries this will be an extremely difficult task to achieve in the immediate future, taking into account the various demands on the financial resources." p. 49.
90. Butler, who had served as Deputy Secretary-General of the ITU since 1968, was elected to the post of Secretary-General at the 1982 ITU Plenipotentiary conference held in Nairobi, Kenya. Some observers attribute Butler's 1982 election to this demonstrated awareness of the telecommunications needs of LDCs who constitute a majority of ITU members. Central to Butler's campaign posture was his active support the GLODOM concept. At numerous UNISPACE '82 "Poster Sessions" (displays or lectures) and press conferences, Butler argued how satellite communications could solve many of the geographical, technical, and financial problems facing LDCs seeking to construct telecommunications systems, particularly into rural areas.

    At the January 1983 Pacific Telecommunications conference in Honolulu, Butler continued to promote the GLODOM proposal as the most feasible way of establishing telecommunications among the island countries of the South Pacific region. Source: Interview and observations by author at UNISPACE '82 and PTC '83. See also, Richard E. Butler, "Telecom Development: A Global Problem Needing World Solution," in *Telephony*, October 24, 1983, p. 9-13 of ITU *Teleclippings*, November 21, 1983.

91. The cost estimates of a GLODOM earth station with a 1-4.5 meter diameter antenna capable of 4 telephone circuits and radio and television reception are placed at about $25,000. In comparison, INTELSAT Standard A antennas (30 meter diameter) and receivers, which are designed to operate with low power INTELSAT satellites, cost about $2 million. The INTELSAT Standard B antennas (10-11 meters) have prices in the tens of thousands of dollars, depending on location and other variables.
92. The cost differential in antennas designed for certain orbital spacing plans is already apparent in the United States. The present 4-degree spacing for C-Band satellites will be gradually reduced to 2-degree spacing by 1990. A private homeowner's antenna, capable of discriminating between satellites two degrees apart, costs between $300 and $500 more than an antenna suitable for 4-degree spacing. Presently there are over 120,000 such antennas in use in the United States alone. The National Cable Television Association estimated the costs to cable operators of meeting the two-degree standards to reach $140-170 million. Interview with sales representatives of Paraeclipse Systems, Santa Barbara, June 2, 1983. See also, Jay C. Lowndes, "FCC Proceeding With Plans for 2-Degree Orbital Spacing," *Aviation Week and Space Technology*, February 11, 1985, pp. 73-75. As one expert told the FCC, "The size of the antenna is the most important factor affecting the complexity and the cost of a television-receive-only installation . . . even though an increase in size of the antenna does not necessarily produce a drastic increase in manufacturing cost, considerations like wind loading, liability and obtrusiveness have major impacts on the complexity of the installation."
93. Quoted at Pacific Telecommunications Conference, Honolulu, January 19, 1983.
94. ITU, "The Missing Link. . ." (1985) Executive Summary, p. 4.
95. Quoted in Dizard, p. 87.
96. Dizard, p. 160.

CHAPTER 2

GETTING A PARKING SLOT IN SPACE

Trying to get a handle on everything that's happening in communications today is like hugging Jell-o. [1]

- *Diana Lady Dougan*

Introduction

One of the unique aspects of the politics of communications satellites is that policy disputes are often disguised under an esoteric technical cloak. For example, the hotly-contested issue of whether direct broadcast satellites may aim their programs to countries who may not want them, is translated to a question concerning "power flux densities" or "sidelobe suppression." This seemingly benign technical gloss tends to deflect laypersons away from the immediate and far-reaching economic and political stakes hinging on the standards being negotiated. So that non-technical readers do not suffer a syntactic lobotomy as we delve deeper into the recesses and folds of the outer space policy-making process, this chapter will review the technologies, concepts, and requirements as part of technical access to the orbit/spectrum resource and satellite communications.

The fact that this technical cloak, which hid ITU telecommunications negotiations from much of the political limelight typical of other international organizations, has been "discovered" is, in itself, a significant indication of the politicization of the orbit/spectrum resource. Knowledge about information technologies is bargaining power. Consequently, negotiators at ITU conferences must comprehend a wide range of technical and engineering technologies, as well as appreciate economic, political, and military ramifications inherent in technical standards and procedures — such as how to measure radio interference. Of course, since the HDCs possess the most advanced technologies, their delegates are also more familiar with engineering and technical parameters that will best promote their countries' interests.

Technical Access

Technical access to the orbit/spectrum resource is accomplished by launching a satellite into the geostationary orbit and communicating with that satellite using radio waves. Briefly, a satellite communications system works as a three-stage process: (1) earth-to-space, called the "uplink"; (2) the signal processing in the satellite itself, designated as the "space segment"; and (3) space-to-earth segment called the "downlink."

Telecommunications traffic, or more simply put — messages — in the form of telephone, television, or other data services, is transmitted as electromagnetic signals, or radio waves, from the "dish" antennas of an earth station to a particular satellite "parked" in the geostationary orbit. The satellite receives the signals through its onboard receiving antennas; the satellite's electronics then re-amplify and re-transmit the signals, usually on alower frequency, back to earth. Earth stations collect the signals in dish antennas and send them to receivers that process and, finally, direct the signals into the terrestrial telecommunications network and on to the end user.

Technical access requires the following: (1) a satellite, (2) a launcher, (3) earth stations, and (4) access to favorable portions of the geostationary orbit and radio spectrum on an interference-free basis. Fundamental differences between these components of technical access focus attention on the orbit/spectrum resource. The first three can be purchased from a growing number of suppliers for prices that decline with economies of scale, maturing technologies, and growing international competition between equipment manufacturers.

It is, in fact, becoming a buyer's market for some of the world's most advanced satellite technologies. While the insurance firm, Lloyd's of London attempts to sell its two salvaged satellites at bargain basement prices, Telecom Canada also has one bird (Anik C-2) hanging in a geostationary storage rack waiting for a buyer. Not just satellites, but rocket firms as well, are trying to pull buyers onto their pad for cut-rate deals to the geostationary orbit. The Space Shuttle is currently flying in the face of growing domestic competition from expendable launch vehicles such as the Delta and Atlas-Centaur, and from Ariane, Japanese, and Chinese launchers abroad.

Ironically, as prices for the technologies decline, the value of the resources will increase, as more users seek access. However, interference-free access to the orbit/spectrum resource cannot be purchased, since it does not result from the actions of any single country. Instead, access to the resources depends upon the cooperation of other users. Since they belong to no one country, they belong to all; they are part of the global commons.

Technical Access and the ITU

The ITU is the intergovernmental organization that allocates, regulates, and manages these two valuable, but limited, global commons resources. The ITU World Administrative Radio Conferences draw up legal guidelines under which a particular user may gain the internationally recognized right to interference-free use. In short, the ITU's mission is to formulate agreements among its more than 158 member states on procedures designed to permit all states the use of the orbit and spectrum resources free from interference. This is the complex task for the ITU's 1985 and 1988 World Administrative Radio Conferences. How should the ITU allocate rights to the orbit/spectrum resource? Should states receive a right upon demonstrating that they are already planning or actually using the resources (the so-called *a posteriori* or "first-come, first-served" approach); or are states entitled to portions of the orbit/spectrum resource solely on the basis of their existence as sovereign states and members of the ITU (the "detailed *a priori* planning" approach) even though they may not use their allotments for years to come?

Avoiding or reducing interference between users is the overriding goal of the ITU resource management rules. Avoidance of interference, in turn, is accomplished through accurate assessments of the capacity of the resources given a particular technology and the number of users. Seen in this way, technical access is a function of the capacity, or scarcity of the resources. Moreover, scarcity is the litmus-test to determine whether the resources are being politicized by countries seeking access or by countries attempting to change the way the resources are used. This chapter will examine resource scarcity from the perspective of technical access; the physical laws governing the orbit/spectrum resource and the technologies that use them.

The Resources

The geostationary orbit and the radio spectrum are paradoxical resources. They are often considered "scarce" within the infinite reaches of outer space. They are designated as "natural," and yet have no substance. They are called "resources," but are non-depletable in the sense they are not consumed after use has ceased; they return to their original condition. They both resemble "locations," one being a flight-path that behaves as a "place"; the other is a region among electromagnetic energy fields that move at the speed of light. They both have incalculable value for political, economic, and military purposes; and yet, they cannot be owned.

The Geostationary Orbit

The special qualities of the geostationary orbit are in many respects a coincidence of physical laws:

> [E]arth satellites orbit about the center of the earth with a period determined by the radius of the orbit. At a radius of some 42,165 km, corresponding to an altitude of 35,787 km, a satellite has a period of 23 hours 56 minutes and is therefore synchronous with the [period of the] earth's rotation [i.e., geo-synchronous]. [2]

This is the family of orbits designated as "geosynchronous" by the ITU, where "an earth satellite['s] period of revolution is equal to the period of rotation of the Earth about its axis." [3] A geosynchronous satellite is found by an earth-bound observer at the same spot in the sky at the same time each day. However, its path, if traced throughout a 24-hour period, would follow a figure eight pattern, the range or size of which is determined by the satellite's orbital inclination.

The actual orbit resource for satellite communications is a subgroup of the geosynchronous family of orbits, where

> [if] the satellite is moving in the same sense as the Earth, from west to east, and the satellite orbit is over the equator, the satellite will appear to an observer on the earth's surface to be stationed at a fixed point in the sky, i.e., geostationary.

The ITU Radio Regulations as ratified international law define a geostationary satellite as:

> A satellite, the circular orbit of which lies in the plane of the Earth's equator and which turns about the polar axis of the Earth in the same direction and with the same period as those of the Earth's rotation. [4]

In much of the literature, the terms "geosynchronous" "geostationary" are used interchangeably. However, as James Gehrig points out, there are legal differences in ITU procedures granting protection from harmful interference depending on whether a satellite's orbital designation is "geosynchronous" or "geostationary." While all geostationary satellites are geosynchronous, the reverse is not true [5]. Operational reality presents a potential definitional dilemma in that satellites designated as "geostationary" are accorded a legal status not extended to "non-geostationary," i.e., geosynchronous or other satellites in earth orbit.

The ITU legal definition of the geostationary satellite orbit (geostationary orbit), although precise, becomes problematical when applied to actual operating space stations exposed to the vagaries of the space environment and their own station-keeping characteristics. In reality, "satellites rarely satisfy the definition." For the following reasons, the dividing line between geostationary and non-geostationary is difficult to draw.

Satellite "drift," away from the definitional ideal of "stationary," is due to the gravitational effects of the oblateness of the earth, moon, sun, and solar radiation [7]. Propulsion systems on board the satellite for "station-keeping" purposes can maintain the satellite's position to within plus or minus .1 degree, "both in longitude and latitude, corresponding to a square 150 kilometers from North to South and 150 kilometers from East to West." The altitude will vary about 30 kilometers [8].

In Gehrig's description, the geostationary orbit can be pictured as an

> annulus-like three-dimensional corridor in which satellites travel at different speeds, altitudes and inclinations to the plane of the earth's equator. [9]

A geostationary satellite's position is expressed in terms of degrees longitude corresponding to the point on the equator over which it appears to hover, a designation that identifies portions of the orbit assigned to satellites (as parking slots).

Up to now, there has been no conflict between the definitions of geostationary and geosynchronous since the distinction is one of degree, not of kind, and because portions of the resource are only now becoming crowded. However, as Gehrig points out,

> Under the [Radio] Regulations a satellite network using a *non-geostationary* [emphasis added] satellite must give way (cease transmission) to a *geostationary* satellite network whenever there is insufficient angular separation and unacceptable interference to the geostationary satellite network operating in accordance with the regulations. [10]

The distinction is blurred because it does not establish at what point, given natural and technically-caused variations in the orbits, a satellite crosses the legally-defined boundary from a geostationary to geosynchronous orbit, thereby losing legal protection. On the other hand, a definition based on "state of the art" station-keeping technologies and technologies and procedures could be prejudicial towards cheaper or older designs, assuming a continual enhancement in technological station-keeping capabilities, accuracies, and satellite lifetimes.

In practice, the definitional differences between geostationary and synchronous satellite orbits has not led to coordination difficulties with the ITU's International Frequency Registration Board (IFRB) procedures. The Radio Regulations stipulate that a geostationary satellite must be capable of maintaining its position to within plus or minus one degree, thereby defining in a legal sense the parameters of "geostationary" as opposed to "geosynchronous." [11]

Technical and Economic Advantages of Geostationary Communications

Having now defined the geostationary orbit, the question inevitably comes up: Why is it such a desirable area of outer space? There are two chief reasons for the usefulness of the geostationary orbit for communications: (1) the apparent motionlessness of the satellite; and (2) its characteristic height above the earth.

Because a satellite in the geostationary orbit appears motionless to an observer on earth, it is also "stationary" with respect to narrow-aperature microwave dish antennas which must accurately "fix" upon the satellite in order to collect sufficient amounts of its weak signals. Because the target satellite does not appreciably move, neither do the earth station antennas, eliminating the need for expensive computer-controlled satellite tracking equipment [12].

The "stationary quality" of the satellite has caused a cost inversion in the design of satellite systems. Due to the fact that the first communications satellites, such as Early Bird, were physically small with only rudimentary station-keeping capabilities and very weak transmitters, the system's major investment was concentrated in the earth segment's large, precisely pointed antenna structures and highly sensitive receivers. Beginning in the early 1970s, communications satellites have become progressively larger, more powerful, and virtually driftless in their orbits, changes allowing the use of much smaller and less expensive earth stations. The savings realized in thousands of simplified earth stations greatly offset the extra expense for more precise station-keeping systems, solar cells, and higher power satellite transmitters. In short, this cost inversion has made satellite networks consisting of thousands of earth stations pathways economically feasible. The chief economic advantage of satellite communications lies in the satellite's ability to interconnect many widely dispersed earth stations with high-capacity data links.

The Connectivity Function

The "connectivity" advantage of the geostationary orbit stems from the satellites' characteristic height of 35,700 kilometers. From a hovering point above the equator a satellite can see, and be seen, from over 40 percent of the earth's surface,

> ... in other words, [from] a circle extending in latitude from 81.3 degrees North to 81.3 degrees South, and in longitude 81.3 degrees West to 81.3 degrees East from the subsatellite point. [13]

As postulated by writer-scientist Arthur C. Clarke in 1945, three satellites in the geostationary orbit can see and interconnect any two communicators between any inhabited areas on earth.

Combined with the "stationary" characteristic, the "connectivity" function projects a revolutionary role for satellite communications in terms of reducing the costs of information distribution. For the first time in the history of communications, the concept of distance has little or no effect on the costs of disseminating information. The cost of satellite communications is distance insensitive.

Cost insensitivity to distance and the connectivity function of satellite communications are a consequence of the fact that once service is established, no additional costs are incurred in beaming data to an additional earth station or communicator within its service area. The satellite can send its signal to all visible receivers, regardless of their location or distance relative to each other. For this reason, satellites offer important techincal and economic advantages over cable or microwave point-to-point relay networks.

For example, a satellite telecommunications network may consist of hundreds of earth stations located thousands of miles apart, and yet they are all interconnected with each other through a common satellite.

For this same capability, a terrestrial network would have to construct $1/2N(N-1)$ links (N=number of communications nodes), the cost of each link increasing with the distance and the amount of materials and resources needed for its construction and maintenance [14].

The Trunking Function

The connectivity characteristics outlined above, which make satellites an especially appropriate technology for linking many dispersed communicators, also allow satellites to act as "cables in the sky." The satellite, by concentrating its beams between two or three high volume traffic gateways, can relay great amounts of data and other forms of traffic for the same distance-insensitive price. This is the pattern of satellite beams found over the North Atlantic, where INTELSAT "birds" link COMSAT teleports in the US with their Western European counterparts.

The Capacity of the Geostationary Orbit

The capacity (or scarcity) of the geostationary orbit is determined by two sets of resource-related factors: physical and spectrum capacity. Physical capacity is the number of satellites that can be placed in the geostationary orbit with a negligible probability of collision. Spectrum capacity refers to the number of satellite communications systems that can operate on the same frequency bands without harmful interference to each other.

Let us take a look at the factor of physical capacity by asking, "just how much space is there in the geostationary orbit region in which to put geostationary satellites?" The region of outer space in which a satellite's flightpath appears "stationary" corresponds to a volume of space of 300 billion cubic kilometers. The circumference of the geostationary region is approximately 265 thousand kilometers, so that one degree of spacing is about 735 kilometers long.

The physical capacity, then, of the geostationary orbit is determined predominately by the station-keeping capabilities of geostationary satellites — their ability to maintain a position and avoid collisions with other space objects in that region of space. Given the present station-keeping accuracies of plus or minus 150 kilometers, approximately 1733 geostationary satellites could be placed in orbit with negligible probability of collisions between controlled satellites [15]. Physical crowding is not considered to be a factor that at present limits use of the orbital resource to an extent warranting its perception as "scarce."

The Earth Station: Center of the Debate

Although the simple and inexpensive earth station assumes a major role in LDC development strategies and in plans for direct satellite broadcasting by both LDCs and HDCs, it also poses costs in terms of "efficient" orbit/spectrum utilization and capacity. In many technical respects, the controversy in the ITU over orbit/spectrum management boils down to the trade-off between orbit/spectrum conservation and the design and cost of earth station antennas. Briefly, the larger and, therefore, more expensive the earth station antenna, the closer together the satellites may be placed in the geostationary orbit using the same frequency bands. If smaller antennas are used, the satellites must be placed further apart in order to avoid interference. Smaller antennas are, technically, less efficient users of the orbit/spectrum resource because they are less capable of discriminating between signals from closely-spaced satellites. In order to more fully analyze the arguments put forward in the debate over technical access, we now turn our attention to the spectrum resource.

The Radio Spectrum

Radio spectrum describes the oscillations of electromagnetic waves that propagate at, or near, the speed of light through space or along conductors. The number of oscillations or waves per second is measured in terms of "frequency" and expressed in honor of their discoverer — Heinrich Hertz (Hz). For radiocommunications purposes, this number ranges from approximately 10,000 Hz (or more commonly designated 10 KHz) to 300,000,000,000 Hz (300 Gigahertz, or 300 GHz) [16]. Radio waves can also be measured as the distance between crests of successive waves, which in the given range of frequencies vary from around 30,000 meters to less than 1 millimeter. There is an inverse relationship between measurements of frequency and wavelength; the higher the frequency the shorter the wavelength, and *vice-versa*.

We can think of the spectrum and its users as a type of community. The spectrum is a very long street sub-divided into lots and houses on each side, all of which are identified by a particular address (frequency). Since a happy household is one composed of compatible residents, the ITU designates a particular house or a particular address as appropriate for a specific type of user-occupant (frequency bands). In this way, spectrum "smokers" and "non-smokers" are prevented from interfering with each other by their ITU community planner.

Although satellite services have been allocated frequencies in many different regions of the spectrum, three frequency bands are in widespread use for commercial satellite communications (designated here in their commonly-used nomenclatures): the C-Band, 6 GHz uplink to the satellite from earth, and 4 GHz downlink, from the satellite to earth stations; Ku-Band, 14 GHz uplink, 12 GHz downlink; and Ka-Band, 30 GHz uplink, 20 GHz downlink (see Table 2.1 below) [17].

(Direct Broadcasting Satellites (DBS) also use a portion of the 12 GHz band for downlink to consumers' home terminals.)

Table 2.1
Primary Commercial Satellite Frequencies and Nomenclature

Band	*Uplink*	*Downlink*
C-Band	6 GHz	4 GHz
Ku-Band	14 GHz	12 GHz
Ka-Band	30 GHz	20 GHz

The spectrum can serve as a communications pathway by intentionally deforming or "modulating" electromagnetic waves. A transmitting antenna radiates the waves into space where they are collected by receiving antennas. The receiving equipment also demodulates, or interprets, the deformations in the waves, thereby extracting the data or information transmitted by the sender.

The spectrum is an ubiquitous phenomenon of nature and is accessible to anyone with even rudimentary technology. As such, the spectrum is a global commons resource due to its universal accessibility and non-excludability. Paradoxically, these characteristics invite interference, the spectrum's chief limitation for universal or unlimited use. To many communicators transmitting simultaneously on the same frequencies will cause "harmful interference." In the case of two communicators generating radio waves on the same frequency, the data in one's electromagnetic waves will be received superimposed upon the other's, resulting in "garbled" or unintelligible data. The spectrum, and, concomitantly, the geostationary orbit are both limited by electromagnetic interference.

For this reason, effective and efficient use of the orbit and spectrum resource requires coordination between users. Due to their mutual need to communicate, states perceive strong incentives compelling cooperation even at some cost to national sovereignty. In a political sense, then, the spectrum forces states to relegate a portion of their spectrum prerogatives as sovereign states to an international regulator and manager of the resource, the ITU. This is the political cost of spectrum scarcity.

The spectrum's scarcity is exacerbated by its non-uniformity or uneven applicability for various types of telecommunications services. Certain portions of the frequency spectrum are more desirable than others. The value of the spectrum resource for relaying information is determined by factors relating to (1) the intrinsic nature of the resource itself, (2) propagation factors related to its environment, and (3) the applicability of available technology.

1. Properties of Spectrum

The nature of the resource itself makes certain portions of the spectrum more cost effective and technically feasible, and, therefore, more desirable.

This reduces the amount of spectrum usable for satellite communications, and heightens competition between satellite and other users of these favorable portions of the radio spectrum.

The information-carrying capacity of the spectrum rises with frequency. The more waves per unit time provide more sites to modulate, or to place message "intelligence." For example, a telephone circuit requires about a four thousand Hertz (4 KHz) "bandwidth" for toll-quality voice transmission. In contrast, a television signal embodying a much greater amount of information in the form of both sound and visual intelligence, requires over a four million Hertz (4 MHz) bandwidth.

As lower frequency bands become congested or saturated, technology is "pushed" by demand for increased capacity to fashion innovations which allow the use of higher frequency bands and expanded information-relaying capabilities. The factor slowing technology's upward drift into new frequency frontiers is the higher energy level of waves in the new bands. The wave must have a lot more energy to be able to oscillate as fast as it does at higher frequencies. This energy also makes the wave harder to hold down onto a wire or inside a transistor. More energetic waves make modulation, amplification, transmission, and, consequently, reception, a more complicated and expensive task for the technology. As we shall see, users are not anxious to abandon lower, cheaper frequencies until they perceive either technical or economic incentives to adopt resource-efficient technological innovations at higher frequencies. Ironically, the countries just now implementing telecommunications services are often the first to use higher frequencies and advanced techniques because they lack decades of investment in older equipment, as is the case in many of the "advanced" countries.

2. Propagation Factors

The propagation properties of radio waves passing through diverse environments such as outer space, the atmosphere, clouds, and rain, vary with frequency. Some frequencies, particularly those above 4 MHz and below 27 MHz, are blocked or reflected by the ionosphere making them less-favorable for satellite use. Above 10 GHz, in the Ku- and Ka-Bands, heavy rainfall can seriously degrade a satellite's signal. The raindrop sizes begin to approximate the wavelength of the signals, making them act as billions of little antennas soaking up the waves' power. Geopolitically, since many developing countries are located in rainy equatorial zones, they see definite national interests at stake in gaining guaranteed access to those frequency bands less affected by rainfall, notably the already crowded C-Band (6/4 GHz). In this way, the North-South debate in the ITU quickly becomes a high-low frequency struggle.

3. The Technology

Technology constitutes the third set of factors determining communications capacity of the spectrum resource. Since crowding is the chief limit to spectrum use, the cost of new, improved technologies for access to higher, less-congested frequency bands becomes a controversial factor influencing technical access to the resources and satellite communications. In other words, who is going to pick up the technology bill? So far, the check is still lying on the negotiating table.

3.1. Dynamics of Innovation

The upwardly spiraling interaction (and interdependency) between the number of spectrum users, applied technology, and cost is set into motion by a change in any one of the three factors. A new, lower cost technological innovation in earth station design brings more users onto the frequency band. This, in turn, creates congestion, requiring a more sophisticated and expensive technology with the capacity to adequately handle the enlarged demand or to expand (conserve) the amount of usable spectrum. In telecommunications, technology creates its own demand.

Past experience has shown that research and development in telecommunications techniques and equipment have expanded the amount of exploitable spectrum by, first, permitting access to higher frequency bands with greater information-relaying capacities, and, second, by making possible the more efficient use or re-use of frequency bands. For example, the current series of INTELSAT V satellites re-use the C-Band frequencies (6/4 GHz) four times with a capacity for over 12,000 telephone circuits. The upcoming generation of INTELSAT VI-series satellites will each carry over 36,000 telephone circuits, re-using both C-Band and Ku-Band (14/12 GHz) frequencies.

3.2. Extension of the Spectrum Street

The perception of spectrum scarcity is primarily premised on projections of future technological expansion of the resource and the costs. Research and development seek to (1) increase the efficiency of exploitation of already utilized frequency bands, and (2) extend the range of usable frequencies. Politically and economically, it is often less costly to extend technology to new, unused frequencies than to attempt to integrate more efficient techniques in frequency bands where there already exists a sizable investment in equipment. This is illustrated by the competition for C-Band frequencies.

The satellite services were born into a harsh spectrum environment marked by crowding and congestion, as large numbers of entrenched users zealously protected the frequencies used by their extensive terrestrial microwave networks. The C-Band was already heavily used by terrestrial and radar services when the ITU allowed the newly-implemented space services to also use portions of the band as outlined by the 1963 Extraordinary Administrative Radio Conference (EARC '63). Since then, of the allocated bands, the C-Band has become the most widely and intensely used for space services, due to its imperviousness to heavy rainfall and its readily available and tested technology. Satellite users, however, must compete with terrestrial services for spectrum "slots" that are particularly scarce in large metropolitan areas. Consequently, large C-Band dish antenna facilities, referred to as "Teleports," are now located either in rural regions or in specially designed locations electronically isolated from terrestrial sources of spectrum "pollution," an isolation that can greatly increase the costs of satellite communications.

The upper-end of spectrum expansion lies within the optical range of frequencies. Laser technologies are being developed and tested for use both with fiber optics and with point-to-point communications between satellites and earth stations. These frequencies enable communicators to relay vast amounts of data measured in "giga-bits," or billions of bits per second. In addition, the very narrow beamwidth of the laser beam enables the sender to focus directly upon the intended receiver, a feature attractive to military users concerned about the beam's security from unwanted listeners. Lasers operating in the blue-green part of the spectrum are being tested as a means for communicating with deep-diving nuclear submarines. Water's blue-green window of transparency may allow geostationary satellites to relay strategic information on laser beams to the submerged submarines.

The major drawback to the use of lasers for satellite communications lies with the large propagation losses incurred as the beam passes through atmospheric obstructions such as clouds or rain. However, in the airless vacuum of outer space, lasers operate unhindered by an obscuring atmosphere. Taking advantage of this fact, future communications satellites and large geostationary platforms will be interconnected via laser beams, the so-called "inter-satellite links" (ISLs).

ISLs are seen as a technological cure for the satellite affliction known as the "double hop." A "double hop" describes the process of completing a circuit by sequentially bouncing the signal twice off satellites. As an hypothetical example, a person in California wishing to call India is not within the service area of the INTELSAT Indian Ocean satellite. The signal from California may be uplinked to one of the INTELSAT Atlantic Ocean satellites, downlinked to an European earth station and then uplinked again to the Indian Ocean satellite, whereupon the satellite downlinks it again to the intended earth stations in India. A double hop requires that the orbit/spectrum resource be utilized two times to connect the two ends of a circuit. In contrast, if the Atlantic Ocean satellite had been able to transmit, via ISL, the telephone conversation directly to the Indian Ocean satellite, orbit/spectrum use (and the signal delay) would have been reduced by half [18].

For many, the technical solution to spectrum crowding is seen in development of satellite systems utilizing the heretofore unused Ka-Band. Many engineers think of the Ka-Band as the "wide open ranges" for future herds of satellites seeking uncrowded, unpolluted, and technologically feasible frequencies. The potential communications capacity represented by the Ka-Band's frequencies is put forth by the HDCs' negotiators, as proof that a technical solution to the expected saturation of lower frequency bands exists. As two satellite telecommunications experts stated:

> Twenty years after Telstar, satellites now provide telephone, television, data, and business services internationally, regionally, and nationally. The geostationary arc has become crowded at C-Band (6/4 GHz) and Ku-Band (14/12 GHz), spurring international plans to use it more efficiently.
> But a host of versatile communication satellites will take to the skies in this decade to add to this crowding. *However, use of the 30/20 GHz (Ka) Band should relieve these orbital-saturation and resulting interference problems now looming at C-Band and Ku-Band.* [19]

Weather is a problem in any uncharted frontier or environment, and rain poses the largest barrier to Ku- and Ka-Band spectrum pioneers. To reiterate, the severe signal degradation incurred as the beam passes through all forms of atmospheric water has spurred efforts to develop higher power transmitters that can compensate in space for rainfall losses on the ground.

The greatest amount of current telecommunications spectrum expansion is taking place in the Ku-Band (14/12 GHz). Although Ku-Band signals suffer degradation some heavy rainfall, for countries outside the tropical equatorial zones this band is the solution to many spectrum crowding problems. As with Ka-Band, the advantages of 14/12 GHz include smaller antennas and less spectrum competition from terrestrial users. This enables operators to place the smaller antennas on the user's premises, obviating the cost of local transmission hops to outlying stations as is often the case in 6/4 GHz communications.

While the high-technology countries continue research and commence operations in the Ku-and Ka-Bands, C-Band still remains the least expensive frequency band for satellite communications and the favored band for developing countries.

Citing these reasons, the Group of 77 (voting bloc of "non-aligned" developing countries) lobbied at UNISPACE '82 for the adoption of a paragraph in the final report recommending that high technology countries gradually shift their satellite operations from the saturated C-Band to higher frequencies.

Although not accepted, the proposal indicates the apprehension of many Third World telecommunications negotiators as to their future difficulties in gaining financially feasible access to C-Band frequencies and orbital slots under current ITU procedures.

3.3. More Efficient Spectrum Use

The chief limit to spectrum use is the amount of interference radio receivers can absorb without losing the desired signal's intelligence. Telecommunications research and development efforts attempt to use spectrum more efficiently by isolating receivers from unwanted electromagnetic "noise." There are three techniques for isolating receivers from harmful interference: spatial, polarity, and temporal isolation.

Spatial isolation can be achieved through three methods: orbital spacing of satellites, service area geographical isolation, and narrow antenna beamwidths. The importance of spatial isolation was underlined by United States telecommunications negotiator, Anthony M. Rutkowski, who called it the "third resource" in conjunction with the orbit and spectrum resources [20].

Perhaps the most politically controversial way to isolate signals from interference is by increasing the spacing between satellites in the geostationary orbit. The earth station antenna "sees" only one source of signals on that frequency for its range of vision, or "beamwidth." While effective, this method also sharply reduces the number of satellites that can simultaneously serve the same service areas on a particular frequency band.

Geographically isolated service areas permit the re-use of frequencies from satellites located in the same section of the orbital arc. With advanced station-keeping techniques satellites can be "collocated," allowing closer spacing and more intense utilization of particularly favorable portions of orbital bandwidth for certain connectivity functions, such as for providing links between Japan and Western Europe, for example.

The "Block Allotment Plan," proposed by the United States for the Broadcasting Satellite Service for Region 2 (North and South America) at the 1983 Regional Administrative Radio Conference (RARC), recognized the advantages posed by the geographical separation of the North and South American continents. This allows the collocation of direct broadcast satellites in essentially the same geostationary position, since they transmit to such widely dispersed service areas as Brazil and Canada.

Spatial isolation may also be achieved through the use of narrow antenna beams, often called "spot beams." Using spot beams, otherwise geographically close service areas can be electromagnetically isolated. Again, spot beams allow the re-use of frequencies by sharply limiting the signal's "footprint" to the intended service area and not allowing significant signal energy to spill over onto adjacent, but unintended service areas.

In sum, spatial isolation allows spectrum re-use through the complex interaction and manipulation of antenna beamwidth characteristics, geographical service area dispersion, and orbital spacing [21].

Polarization isolation permits the re-use of frequencies by generating and filtering the signals according to their polarity. To put it as succinctly as possible, a radio signal or beam consists of electrical and magnetic fields of various polarities. However, a particular polarity may be produced through the mechanical design and orientation of the antenna elements. With proper adjustments "vertical" and "horizontal" beams may be simultaneously transmitted or received independently of each other on the same frequency. For example, the receiving antenna can discriminate between polarities so that it receives only the horizontally polarized beam. In this way, the capacity of the available bandwidth is doubled.

Polarizations may also be circular, designated as either "right hand" or "left hand" circular polarizations. Unfortunately, beam polarizations are sensitive to changes in the propagational environment. Signal degradation in the higher Ku- and Ka-Bands due to precipitation also suffers from the signal's polarization. Heavy rainfall changes the signal so that the receiving antenna sees a distorted polarization of diminished strength. The C-Band frequencies are less susceptable to atmospherically-induced polarization changes, an aspect which again emphasizes the C-Band's attractiveness to satellite operators.

Satellites' use of frequencies and service areas can also be isolated temporally so as to limit interference. Transponders may be overloaded by too many earth stations demanding circuits at the same time, particularly when a satellite is performing a connectivity function. This type of interference is most acute in the "multiplexing" function of satellites. Multiplexing techniques enable the satellite to interconnect many different communicators by allowing the earth stations to select the intended recipient earth station-by circuit.

Conventional techniques for multiplexing utilize *analog* frequency-modulated (FM) signals, where each circuit, or interconnection, was assigned a specific frequency within a particular satellite transponder. Analog signals as utilized in conjunction with frequency-division multiple access (FDMA) present limitations in total frequency use due to satellite:

> ...[a]mplifier nonlinearities [that] generate intermodulation products from multiple carriers. This effect at the satellite transponder can be reduced, though, not eliminated, by "backing off" the drive level to the amplifier, which in effect reduces the received carrier level and increases the effect of thermal noise on signal quality at the earth receiver [thus requiring larger, more expensive earth station antennas]. [22]

Digital signal multiplexing techniques have expanded the communications and connectivity capacity of spectrum due to the computer's ability to manipulate a digital transmission. Digital signals consist of "bits" of information that exist only as "on" or "off" impulses of electromagnetic energy, corresponding to the binary (1 or 0) machine language. While frequency-division multiple access isolates discrete signals by frequency, digital techniques allow stations to be isolated by timed bursts of digital data.

By computer-ordered sequential timing among all earth stations communicating through a satellite's transponder, each sender can fully occupy the entire bandwidth and transmitter power of the transponder. Within a 750 microsecond time frame, each communicator fires a burst of data encoded with the "address" of the desired receiver. The sequence may vary according to traffic patterns and demand, but each communicator's transmissions are sent at distinct *times* when no one else is using the frequency, reducing interference. This technique is called time-division multiple access (TDMA) [23].

TDMA offers greater capacity and reduced interference for satellite networks and the orbit/spectrum resource, but this is accomplished with an increase in complexity and costs. Each earth station must be precisely and accurately synchronized with all other earth stations and the satellite. This technique is most practicable for high volume traffic networks with skilled maintenance personnel.

Technical Access to the Combined Orbit Spectrum Resource

The preceding sections have briefly outlined the geostationary orbit and radio frequency spectrum as discrete natural phenomena and resources. This section will consider the combined orbit/spectrum resource as utilized by communications satellites. The purpose of this section is to assess the communications capacity of the orbit/spectrum resource. In doing so, this section looks at the issue of orbit/spectrum resource "scarcity" from several different perspectives. A perception of resource scarcity is, in many respects, the best predictor for countries' policy stances towards ITU negotiations on orbit/-spectrum resource allocations and planning.

Characteristics of the Combined Orbit/Spectrum Resource

The present physical capacity of the orbit/spectrum resource is not appreciably limited by the risk of collision between satellites or other objects and debris in the geostationary orbit region [24].

The present low risk of physical collision shifts the focus of attention to spectrum crowding as the chief limitation to orbit/spectrum capacity.

Discrete Orbit and Spectrum Resources Defined by Frequency and Service

Interference occurs when communicators seek to use the same frequencies at the same time and in the same service areas. For this reason, the orbit/spectrum resource can be conceptualized as discrete resources, defined and separated by frequency band.

The frequency bands allocated by ITU World and Regional Administrative Radio Conferences (WARCs and RARCs) to satellite services, in effect delineate distinct and discrete orbit and spectrum resources.

1. Fixed Satellite Service (FSS)

As stated above, the frequency bands currently used for the Fixed Satellite Service (FSS) are the C-, Ku-, and Ka-Bands, 6/4, 14/11, and 30/20 GHz, respectively [25].

2. Broadcasting Satellite Service

The Broadcasting Satellite Service (BSS) was allocated uplink frequencies for feeder links in the 17 GHz band for Region 2 countries, and lower frequencies for Regions 1 and 3. The broadcasting downlink would use 12 GHz downlink frequencies in the Ku-Band, the actual frequencies varying by ITU region [26].

Frequencies Allocated to Other Services

1. The Mobile Satellite Services

Communications satellites are being used increasingly for mobile radiocommuniations, designated as the Mobile Satellite Services (MSS). The MSS provide circuits between communicators who are operating "mobile" or non-fixed earth stations on air, sea, or land vehicles. Technical improvements in mobile earth station receivers, transmitters, and stable shipboard antenna platforms have extended satellite communications to ships through the INMARSAT satellite network.

In addition to conventional telephony, maritime satellite services include search and rescue communications between land and ship based computers. The provision of mobile telecommunications is the area of greatest potential growth for satellite services. This future was well illustrated by the pioneering Geostar concept:

> Geostar solves one of the most fundamental problems in highly mobile society: how to find and establish two-way communications with someone wherever he is . . . For most people, the mind boggling thing about Geostar is the notion that a calculator size, hand-held transceiver powered by AA penlight batteries can both receive from and transmit directly to a satellite.
>
> Geostar is able to perform seven functions: (1) positioning, (2) directional guidance, (3) collision and terrain avoidance for boats and

aircarft, (4) position reporting to fleet dispatch headquarters, (5) send messages, (6) receive messages, and (7) interconnection to the telephone system for all database services and other services that can be provided through a modem. [27]

The allocated Mobile Satellite Service frequencies now being used are in the L- and X-Bands, 1.6 and 8/7 GHz, respectively. The special characteristics of mobile satellite communications prompted allocation of frequency bands for their exclusive use. An increasing number of satellites, however, now operate both within and between several different bands, and even interconnect different bands and services, such as with the Tracking and Data Relay Satellite (TDRS) which is used to provide fixed, mobile, and inter-satellite service; the INMARSAT payloads on the INTELSAT-V-MCS satellites that provide ship-to-ship and ship-to-land telecommunications; and the INSAT satellites that perform weather, direct broadcast, and telephony functions from the same space vehicle.

Hybrid satellites are rapidly blurring the ITU distinctions between services and frequency bands. These satellites complicate orbit/spectrum resource management and allocation rules, providing a highly-visible dilemma for proponents of detailed planning of the resources.

2. Military Services

The major users of mobile satellite services are the military forces of the superpowers and those of their alliance members. High-powered satellites enable troops in the field, equipped with lightweight backpack units, to be in constant radio-communication contact with both field and governmental headquarters. Milstar will be such a "hybrid," designed for both mobile and fixed satellite services. It will transmit on several frequency bands as part of the overall C^3 military tactical and strategic command structure.

Military satellite services currently use frequencies in the UHF range, from 235-399 MHz, and in the X-Band, 8/7 GHz. Even though the X-Band is for both FSS and MSS, by international custom, military services use the 8/7 GHz frequency band exclusively.

Assessing Capacity

To assess the communications capacity of the orbit/spectrum resource is an extremely complex and problematical task. Any assessment must take into account a wide and constantly changing assortment of diverse technologies, communications (modulation and multiplexing) techniques, earth station design, atmospheric conditions, and propagation factors, among others. In addition, a continuously changing cost factor associated with evolving technologies hinders general agreement on the eventual capacity or even on a method for its calculation. In fact, it is the negotiating position of the United States that to even attempt to estimate the orbit/spectrum resource's ultimate capacity is wasted effort because technological advances soon render obsolete any such static assessment [28].

Not surprisingly, the ITU has also been unable to arrive at a definition of capacity:

> [A] precise definition of orbit capacity depends on *assumptions* [which are listed, including] performance characteristics of the satellites and ground stations [Emphasis added]. [29]

Efficiency of Resource Use

The difficulty in defining and calculating the maximum capacity is that the assessment serves as a benchmark for determining the "efficiency" of resource utilization [30].

"Efficiency" is a major point of contention between HDSs and LDCs supporting rival proposals at the 1985 and 1988 Space WARC on how to allocate and manage the orbit/spectrum resource. "Capacity" refers to "how many or how much," while "efficiency" raises the more political questions, "capacity at what price, in lieu of what services?" Capacity and efficiency of orbit/spectrum resource use are concepts loosely joined in an ends-means configuration. The tie between capacity and efficiency taints capacity with a political coloring that obscures a more objective assessment. Indeed, the politicization of the orbit/spectrum resource stems largely from differing perceptions of precisely how closely tied efficiency of resource utilization is to its potential capacity.

The following paragraph quoted from a UNISPACE '82 Background Paper on the Geostationary Orbit succinctly outlines the debate over orbit/spectrum resource capacity and efficiency:

> Since 1970 the number of geostationary satellites has increased an average of 18 percent per year. This corresponds closely to the average growth of INTELSAT's telephone circuits which double every four years. If this trend were to continue the geostationary arc would be quickly filled. However, through advances in frequency re-use, antenna patterns, and multi-band capability, the rate of increase in the number of satellites should level off as the capacity of individual satellites is enhanced by more *efficient* technology [emphasis added]. [31]

Estimating Capacity through Assessments of Use

Capacity can be measured as the total use of the orbit/spectrum resource. More precisely, capacity refers to the number of satellites in the geostationary orbit, the amount of spectrum or bandwidth used, earth stations, and communications pathways provided.

1. The Number of Geostationary Satellites

An accurate appraisal of the total number of geostationary satellites is difficult to make for several technical and administrative reasons. First, the number is constantly changing. The population of satellites in the geostationary orbit has increased an average of almost 20 percent per year since 1970. However, this number or rate of increase is affected by the number of actually functioning satellites which fluctuates with the number of "dying" and dead drifting satellites. Once satellites exhaust their supply of station-keeping propellant or suffer other technical malfunctions, they begin to drift in an uncontrolled fashion in, through, and about the region of the geostationary orbit until they come to their ultimate rest in a gravitational graveyard, located, appropriately enough, over the Sri Lanka residence of their originator, Arthur C. Clarke.

The task of counting satellites is complicated further by the different criteria used by various agencies for that purpose. The ITU, the official international regulator and registrar of geostationary satellites, does not independently monitor or track the number of functioning geostationary satellites, nor does it have the technical capacity to do so. Instead, the ITU has the assigned duty to record *notifications* sent to it (actually the International Frequency Registration Board (IFRB)) by administrations voluntarily complying with the ITU Radio Regulations.

Informed observers estimate that as of early-1985 there were between 110 and 115 functioning satellites in the geostationary orbit. The fact that this number is far below the ITU's (approximately 240) has not only technical, but also economic and political ramifications. First, the ITU's figures appear higher for reason of its administrative functions as coordinator and regulator of the resource. Given satellites' five-to-seven year "bid-to-launch" time-frame, the ITU and the IFRB must know years in advance where countries intend to deploy satellites. Coordination procedures may be needed in cases involving *projected* levels of harmful interference between both functioning satellites and those planned for operation in that sector of the orbital arc. This, however, has an economic and political fallout in that costly coordination procedures may be initiated solely on the basis of notified intentions, with the result that assessments of interference are arbitrarily established more by engineering theory and less by actual practice. In sum, the ITU's administrative procedures for assessing the orbit's population and interference-triggering levels for coordination procedures project enormous economic and political stakes onto the engineering models and formulae.

There is, however, one internationally-recognized count of the number of geostationary satellites. At UNISPACE '82, an interim working group, established after the full committee was unable to arrive at a consensus regarding Paragraph 60 of the Draft Report, debated the question of how to assess the actual number of geostationary satellites. Its discussions are a vivid example of showing how different sides differ in their perspectives regarding orbit/spectrum scarcity.

At the second meeting of the interim working group, the Indonesian chairman asked the ITU for their latest count. This was delivered by (then) Deputy Secretary-General Richard Butler on August 13, 1982:

> The total number of satellites operating or planned for operation in the geostationary orbit and notified to the IFRB of the ITU at the 12 of August 1982 in accordance with radio regulations are 248 [sic]. [32]

Butler's report then subdivided the figure of 248 into categories of 176 satellites either operating or having completed coordination, and 72 satellites in "various stages of coordination with the IFRB."

The Indian delegate, Mr. A. Srirangan, preferred the ITU figures as assessed on December 31, 1981, and later published in the ITU's yearly report to the Committee on the Peaceful Uses of Outer Space (COPUOS). The report showed 220 geostationary satellites, of which 162 were operating or having completed coordination, and 58 satellites in the first stages of notification [33].

The agreed-upon wording that was ultimately adopted by both the full committee and the Plenary Assembly states the following international consensus regarding the number of geostationary satellites:

> As of 31 December 1981, a total of about 220 satellites for various telecommunications purposes (including broadcasting, meteorology and other services, as well as experimental) were already in operation or notified to the ITU as planned for operation in the geostationary orbit. Of these, about 63 satellites are for international public telecommunications services (INTELSAT, Intersputnik, and INMARSAT). Of the approximately 157 remaining satellites, about 128 stand notified by developed countries and about 24 by developing countries. [34]

The intent behind the paragraph as proposed by the Indian and Indonesian delegations was to show the relative distribution of geostationary satellites between the HDCs and the LDCs. Approximately four-fifths of the total number of satellites were notified by the HDCs while about one-fifth were reported by the LDCs.

However, the tactic may not have achieved the effect intended by its LDC proponents. Some observers noted that for "such an advanced technology, one-fifth is quite high for developing countries to have." Or, as seen from a slightly more political perspective, the fact that the LDCs use or will use one-fifth of all communications satellites, points out the importance Third World governments attach to telecommunications, and communications satellites in particular.

2. Distribution of Satellites by Services

The UNISPACE '82 figures provide a useful benchmark from which we may evaluate relative distribution of satellites by frequency band. Of the approximately 220 operating, notified, or coordinated geostationary satellites, some 140, or about 63 percent, use the C-Band frequencies. 31 of these satellites also utilize other frequency bands, notably the INTELSAT V series of spacecraft with transponders for Ku-Band trunking and INTELSAT-V-MCS which provide transponders for INMARSAT maritime satellite communications in the L-Band, 1.6/1.5 GHz.

Approximately 53 satellites operate or will operate in the Ku-Band. Japan was the first administration to notify the ITU of its intention to operate three satellites in the Ka-Band, at 30/20 GHz. The majority of the notified satellites for the Ku-Band are direct broadcast satellites operating in the BSS frequency bands. The remainder are primarily military satellites using the UHF, L- and X-Band frequencies.

3. Factors Affecting Distribution of Satellites in the Geostationary Orbit

Satellites are placed where they can "see" the most traffic [35]. Orbit/spectrum capacity is determined to a great degree by the geographical distribution of telecommunications traffic and service areas, particularly for satellites performing a connectivity function. For INTELSAT, there are relatively few sections of the geostationary arc that are visible to an optimal number of earth stations and to the bulk of international telecommunications traffic. The hemispheric dispersion of service areas and earth stations drastically reduces flexibility in satellite placement in sections of the geostationary arc visible to the most traffic over the Atlantic Ocean, Africa, and North America.

The United States and Canadian situations are instructive. With four-degree spacing, 22 C-Band satellites could be parked in approximately 90 degrees of arc with optimal coverage of the continental United States. There are only 25 degrees of arc from which a satellite can serve all 50 states. The same constraints apply to Canadian satellites.

The growth in telecommunications needs and services provided by satellites for the United States and Canadian markets has filled the available orbital slots in the C-Band. With the present four-degree spacing the orbit/spectrum resource is "saturated" in the C- and Ku-Bands. According to a NASA study:

> Demand for satellite communications service in the US will saturate available geosynchronous arc and frequency spectrum capacity in the early to mid-1990s, even if orbital spacing is reduced to 2 deg. in both C- and Ku-Band . . . [36]

In short, the larger the service area, the smaller the optimal portions of the geostationary arc, and the more critical the satellite spacing. There are different orbit/spectrum constraints encountered when satellites *broadcast* their signals to small subscriber earth stations, even within a relatively limited service area. For example, commercial BSS systems in the Ku-Band are economically viable only when used with small, *inexpensive* .8 meter rooftop antennas. However, these consumer-oriented antennas, in turn, require at least six-degree satellite spacing due to the antennas' lower discriminatory capabilities and the high-power satellite transmitters necessary for satisfactory home reception. In contrast, the increased discriminatory power of the same .8 meter antenna used with the shorter wavelengths of the Ka-Band frequencies allow 30/20 GHz satellites to be located within one-degree of one another [37].

Estimating Total Capacity of the Orbit/Spectrum Resource

The most direct method for assessing the capacity of the orbit/spectrum resource is to calculate the total amount of telecommunications traffic carried by geostationary satellite transponders as depicted in Table 2.2 below:

Table 2.2
Total Satellite Traffic: INTELSAT, Regional, and Domestic
Unit: Equivalent 36 MHz Transponder

	YEAR			
	1980	1985	1990	1995
Traffic				
Voice	256	705	1,316	2,484
Data	---	176	440	1,096
Television	55	225	503	1,135
Teleconferencing	---	70	130	249
Total	331	1,176	2,389	4,964

Source: Robert R. Lovell and C. Louis Cuccia, "A New Wave of Communications Satellites," *Aerospace America,* March 1984, p. 44.

While Table 2.2 shows how traffic demand affects assessments of orbit/spectrum capacity, it does not indicate how many satellites or orbital slots will be required to supply the telecommunications services demanded. In order to do so, we must take a much closer look at the technological and methodological bases of the capacity criteria.

As stated in the CCIR IWP 4/1 study, capacity can be assessed only upon many assumptions of antenna patterns, modulation and multiplexing techniques, and satellite performance, especially regarding frequency re-use. The

UNISPACE '82 Background Paper underlines these methodological difficulties while pointing out external factors as well:

> The problem of assessing and predicting the degree of saturation of the geostationary orbit is complex and cannot be solved reliably and quantitatively. The evolution of the demand and supply of services depends on general economic growth, growth of telecommunications, technological advances, costs of existing and new technologies and other factors. Furthermore, these factors interact with each other to make predictions even more hazardous. The interactions are listed as:
> a. As demand grows and as existing systems become saturated, new technology will be developed;
> b. If the new technology requires a rapid increase in price, the growth in demand will be slowed;
> c. New technology for terrestrial communications, for example, fiber optics, may make terrestrial channels more competitive, thereby reducing the demand for satellite services;
> d. Efforts to conserve energy may include replacing travel by communications, creating, for example, a demand for video conferencing by satellite;
> e. Advances in space transportation systems may make possible large space platforms which may provide economies of scale. [38]

The dissimilar parameters of capacity, demand, and efficiency of resource use are major stumbling blocks to acceptance of any assessment. Writing in 1976, Gehrig states:

> There is no agreement . . . on how the use efficiency of the geostationary orbit should be measured . . . For the present, it has been decided that the efficiency criteria will be equal to capacity per satellite divided by the bandwidth times the orbital arc occupancy. There are major objections to this simple expression for measuring efficiency since the relation between orbit occupancy and spectrum utilization is too intricately connected. [39]

Efficiency also shows an economic and resource trade-off. For example, at the Pacific Telecommunications conference in January 1983, INTELSAT presented a plan for providing satellite telecommunications to isolated South Pacific islands utilizing unused capacity on INTELSAT satellites.

In order to establish thin-route rural telephony in the South Pacific island countries at an affordable price, small, low-cost earth stations would be used. The INTELSAT proposal (forerunner to their VISTA Service) envisaged 3-4 meter antennas providing up to four telephone circuits via INTELSAT satellites. However, to interconnect the small "Basic User Earth Stations" (BUES) would require double-hopping the signal from a BUES to a larger Standard B earth station. The switching equipment of the large earth station would route the signal to the frequency the desired BUES was listening to, then send the signal back up to the satellite which would transmit the signal to the intended BUES.

In effect, the small earth station requires that the orbit/spectrum resource be used twice to achieve one telephone circuit. However, in the cost/benefit perspective of a Pacific island country without a reliable or universal telecommunications network, "efficiency" is seen primarily as the number of BUES and telephone circuits established for a given price, and less by the amount of orbital bandwidth used [40].

A calculation of resource use efficiency must take into account two parameters — cost of the technology and the cost in terms of orbit/spectrum resource "waste" or "conservation." A fuller appreciation of the efficiency-conservation debate can be gained by looking at a crowded and uncrowded resource.

1. The Crowded Orbit/Spectrum Resource

Resource conservation is expensive in economic and technological terms. From the user's point of view, resource costs are perceived or realized only under conditions of resource scarcity — when one user pre-empts another — a condition that introduces an opportunity cost to any choice in allocating access to the resources. In such a case, there is an immediate resource (opportunity) cost when a less-efficient user pre-empts a more-efficient user. In other words, efficiency is determined by the degree of saturation or crowding found in that portion of the orbital arc. For example, a satellite that re-used 500 MHz of available bandwidth six times would be twice as efficient as a satellite that only re-used the same amount of spectrum three times.

2. The Uncrowded Orbit/Spectrum Resource

In contrast to the previous exampole, in the case of an uncrowded resource, one user does not pre-empt another. There is no perceived opportunity cost incurred in allowing a less efficient user to use the resources since that satellite still represents an added increment of usable orbital bandwidth which would not otherwise be available.

Ends-Means Dilemma

When capacity and efficiency of orbit/spectrum use are considered in terms of the "ends" of the telecommunications system, calculation or assessment of the two parameters as "means" is made vastly more complicated. If the satellite system is intended to provide connectivity between widely dispersed low-volume users, the cost of the earth stations becomes a crucial factor in the overall feasibility of the network. While efficiency in terms of the number of circuits will be low, the efficiency measured as the number of pathways between communicators will be high.

In contrast, a satellite system designed for high-volume trunking between few "gateway" earth stations is highly efficient in terms of circuits, but not so in its provision of pathways. LDCs desiring to establish low-volume rural networks see orbit/spectrum efficiency measured in the number of pathways, while the HDCs prefer to emphasize the number of circuits provided as the main criterion.

Technology is a changing variable in any assessment or estimate of capacity or efficiency of resource use. Under the present "first come, first served" system, flexibility of allocations is reflected and recognized by the diversity of capacity assessments. As one spectrum analyst wrote:

> In some planning approaches that avoid detailed planning, the capacity need not be defined in detail. Since the resource is not completely divided up, there is no need to determine exactly what it is. In fact, in a flexible approach to planning, the capacity itself remains flexible since the approach is based on the assumption of an *expanding* [emphasis added] capacity. [41]

In summary, there is a perception of scarcity of the orbit/spectrum resource. However, it is not based solely on technical parameters, which, most would agree, point to expanding resource and technological capacities. Instead, the various contending proposals for managing resource allocation and use are, to a large degree, politicized due to the lack of agreement on how to assess capacity and efficiency of orbit/spectrum utilization. However, these only indicate that the politicization has other causes, which will be covered in subsequent chapters.

REFERENCES

1. Diana Lady Dougan, US Coordinator for International Communication and Information Policy, quoted in Special Advertising Supplement, *Time,* October 1983, p. 16.
2. Background Paper presented at UNISPACE '82, "Efficient Uses of the Geostationary Orbit," A/Conf.101/BP/7, p. 5.
3. ITU, Final Acts, WARC-ST, 1971
4. ITU, Final Acts, WARC-ST; see James J. Gehrig, "The Geostationary Orbit — Technology and Law."
5. Gehrig, p. 267.
6. *Ibid.*
7. The sun and moon exert gravitational pressure on the orbital path causing the satellite to incline its orbital plane by .85 degrees per year. Solar radiation can influence the path exerting a slight pressure, the strength of which varies with the area of exposed satellite surfaces. The major external influence is the ellipticity of the mass of the Earth; it can

influence the altitude of the satellite by plus or minus 34 kilometers.

8. UNISPACE '82 Background Paper, "Efficient Uses of the Geostationary Orbit," Vienna, August 1982.
9. Gehrig, p. 267.
10. Gehrig, quoted from WARC-ST Final Acts, Regulation 470VA, Spa2:

    Non-geostationary space stations in the fixed satellite service shall cease or reduce to a negligible level radio emissions, and their associated earth stations shall not transmit to them whenever there is insufficient angular separation between the nongeostationary satellite and geostationary satellites and unacceptable interference to geostationary satellite space systems operating in accordance with these Regulations.

11. Personal interview with Thomas Tycz, Office of Science and Technology, Federal Communications Commission, Washington, DC, December 3, 1982. Mr. Tycz emphasized that satellites are capable of station-keeping far within the limits set by the ITU Radio Regulations.

    Radio Regulation 470VC Spa2: Space stations on geostationary satellites — shall have the capability by maintaining their positions with plus or minus 1 degree of the longitude . . . but efforts should be made to maintain their position at least within plus or minus .5 degree.

12. Additional advantages of the "stationary" satellite include the absence of the Doppler frequency shift associated with moving objects. The doppler shift complicates multiplexing and high data rate transmissions between computers. The stationary satellite also offers enhanced communications reliability because it is continuously within view of the communicators.
13. ITU, Background Paper on Efficient Uses of the Geostationary Orbit for UNISPACE, p. 5.
14. INTELSAT calculates that each satellite communicates between 100 different international and transoceanic pathways. In order to accomplish the same connectivity capability, there would have to be ½ (100) (99)–4,950 terrestrial and submarine cable links. Statement by Dr. Joseph Pelton, Executive Assistant to the Director-General of INTELSAT, at Pacific Telecommunications Conference, Honolulu, January 19, 1983.
15. L. Perek, "Physics, Uses, and Regulation of the Geostationary Orbit, or *Ex Post Sequitur Lex,*" *20th Colloquium on the Law of Outer Space* (1977):400-411.
16. This is the portion of the entire electromagnetic spectrum that is officially allocated and regulated by the ITU. See, Charles Lee Jackson, "The Allocations of the Radio Spectrum," *Scientific American*, (February 1980), pp. 30-35; and Vernon Pandorin, "Protecting Radio Win-

dows for Astronomy," *Sky and Telescope*, April 1981, pp. 308-310.

17. Of course, this is a much-simplified delineation of the actual frequencies, services, *et cetera*. Readers wishing a more precise table of frequencies are encouraged to refer to *Satellite Communications Magazine*, December 1984, centerfold insert. The ITU has allocated other bands to the FSS and other satellite services both higher and lower in frequency than the D-, Ku-, and Ka-Bands described here. Owing to the absence of extensive use or crowding on these bands, they will not be treated in this study other than to mention their existence at this point.
18. See, Anthony M. Rutkowski, "The Inter-Satellite Service: Future Shock for Telecommunication Policy Makers," *Telecommunications*, July 1981, pp. 64-65.
19. Robert R. Lovell and C. Louis Cuccia, "A New Wave of Communications Satellites," *Aerospace America*, March 1984, pp. 43-51, at 43; and Robert R. Lovell, "Giant Step for Communications Satellite Technology," *Aerospace America*, March 1984, pp. 54-57. The United States, Japan, and the European Space Agency are actively testing the Ka-Band (30/20 GHz) for communications dependability under varying propagational conditions. See, US Congress, Office of Technology Assessment, *Civilian Space Policy and Applications*, June 1982, pp. 47-53; and for a description of the Japanese, French, British, and West German space research efforts, pp. 175-214.
20. Personal interview at Forum '83, Geneva, October 30, 1983.
21. Statement by Edward Jacobs, International Office, FCC, at Satellite Users Conference, August, 1982; and personal interview in Geneva, Switzerland, December 10, 1982. Harold G. Kimball, Director of Communications and Data Systems Division, Office of Space Tracking and Data Systems, NASA, in a press conference at UNISPACE '82, outlined the spectrum expansion techniques. the emphasis of the US presentation was on the "unlimited" nature of the orbit/spectrum resource due to continual improvements and innovations in technology.
22. Burton Edelson, *et al.*, "Greater Message Capacity for Satellites," IEEE *Spectrum*, March 1982, p. 58.
23. Edelson, *et al.*, p. 58.
24. A growing number of experts warn, though, that debris and "dead" satellites could pose dangers in the future:

    > Since most satellites are able to maintain their positions with ±0.1 degree of longitude, there are 1800 slots, each 0.2 degree wide, in the geostationary orbit such that there would be no risk of collision between functioning satellites. If two or more satellites are positioned at the same nominal position, there is a risk of collision

which depends on the size of the satellites. A recent study concludes that if 10 satellites were assigned the same location, or "collocated," there would be an average of one collision every 400,000 years between *active* satellites.

Source: Background paper, "Efficient Use of the Geostationary Orbit," UNISPACE '82, A/Conf.101/BP/7, 12-13.

The estimate, however, does not take into account the collision probability between active and non-functioning satellites that are no longer under an operator's control. The argument, made in various UNISPACE '82 committee meetings, states that operators should be required to remove their satellites at the end of their operational lives to other orbits, allowing them to drift away from the flightpaths of functioning geostationary satellites. Although not a serious problem today, the fact that debris in the geostationary altitude will orbit for billions of years means that the problem could persist, literally, for as long as the earth rotates.

25. Donald M. Jansky, *World Atlas of Satellites*. Dedham: Artech House, Inc., 1983, pp. 9-44. Jansky does an excellent review of the technical parameters associated with the various satellite services.
26. *Ibid.*, pp. 129-148.
27. Gerald O'Neill, "The Geostar Satellite System," *Satellite Communications*, March 1984, p. 64. See also, Jansky, p. 105.
28. This was repeatedly stated in many fora and by several United States delegates, but most eloquently in an interview with Tom Tycz of the Federal Communications Commission, December 3, 1982, in Washington, DC.
29. CCIR IWP 4/1, "Provisional Technical Report for WARC 8[5]," p. 172. Document 4/286-E.
30. *Ibid.*, The document continues:

    In the case of dealing with the orbital capacity the definition of a maximum capacity will become an important method to evaluate the geostationary orbit capacity efficiency.
31. UNISPACE '82,"Efficient Uses...," p.19.
32. Statement recorded by the author at UNISPACE, August 13, 1982.
33. ITU, 21st Report on Telecommunications and the Peaceful Uses of Outer Space, ITU, Geneva, 1982, pp. 22-32.
34. UNISPACE '82, *Report of the Second United Nations Conference on the Exploration and Peaceful Uses of Outer Space*, Vienna, 9-21 August 1982. UN Document A/Conf.101/10.
35. See, Henry Chasia, INTELSAT Planning Staff, "INTELSAT's Utilization Orbit, Spectrum and Technology to Meet System Requirements in teh 1990s," *AIAA 9th Communications Satellite Conference*, San Diego, California, March 1982.

36. Source: "NASA Forecasts SATCOM Capacity Saturation," *Aviation Week and Space Technology*, April 18, 1983.
37. There is a very complex economic and technological trade-off between orbital spacing, rain attenuation, antenna size, and satellite transmitting power. For a more complete discussion, see Wilbur L. Pritchard, "High vs. Low Power for Direct-Broadcast Satellites," *Aerospace America*, March 1984, pp. 58-59.
38. UNISPACE '82 Background Paper, "Efficient Uses . . . ." p. 21.
39. Gehrig, Footnote 25, p. 276.
40. Joseph Pelton, *et al.*, "INTELSAT: The Global Telecommunicatons Network," Monograph distributed at Pacific Telecommunications Conference, Honolulu, January 16-19, 1983. In statements made in various conference sessions, Dr. Pelton estimated the cost of the BUES to be "in the neighborhood of $25 to $40 thousand," as compared to about $500,000 for an INTELSAT-compatible Standard B earth station capable of "single hopping" circuits between earth stations.
41. Peter H. Sawitz, "Planning Satellite Communication Services and Spectrum-Orbit Utilization," in *9th AIAA Satellite Communications Technical Papers*, presented in San Diego, March 1982, pp. 489-494.

CHAPTER 3

MANAGING THE "HIGH GROUND": SQUATTERS' RIGHTS AND NEWCOMERS' CLAIMS TO GEOSTATIONARY GOLD

> Periodically throughout history mankind has had the opportunity to develop legal theories to meet the demands of advanced technologies. But advanced technology has only rarely produced whole new arenas for human interaction where no laws govern the interaction. Space is one such arena. [1]
>
> \- *S. Neil Hosenball*

Introduction

The politicization of access to the geostationary orbit and spectrum resources stems from the negotiating "push-pull" between the countries with "technical access" to the resources, and the non-space powers, to create an international legal structure controlling use and guaranteeing access to the resources and technology in a way that furthers a particular state's interests. Establishing such a resource management regime is becoming a source of global competition and conflict owing to the inherent tension between the legal norms of the Outer Space Treaty, which stipulates free and open use, and the increasingly valuable and, therefore, scarce orbit/spectrum resource. This chapter will examine the Outer Space Treaty [2] and introduce the International Telecommunication Union as the foundation building blocks for the emerging legal regime managing the orbit/spectrum resource.

This first section will focus on the central controversy surrounding the legal status of the geostationary orbit: are the provisions of the Outer Space Treaty stipulating free and open exploration and use applicable in practice to the "scarce" geostationary orbit, where the use of an orbital slot by one state may pre-empt its use by another? In other words, are there any "squatter's rights" for the countries already using orbital slots, or do newcomers have supportable claims to those slots?

The Outer Space Treaty and the Legal Status of the Geostationary Orbit

Our analysis here will review the various arguments of legal scholars as they have sought to apply the treaty provisions regarding "use" of the outer space environment to the concept of "access" to a limited resource. Articles I and II of the Outer Space Treaty suggest three principles governing access to the geostationary orbit: (1) free and open access, (2) non-appropriation, and (3) access for the "benefit and province of mankind."

The Outer Space Treaty constitutes the legal basis for extending the rule of law into outer space. Consisting of a preamble and thirteen articles dealing with activities in space, the Treaty annunciates principles governing the exploration and use of the space environment.

However, the legal status of the geostationary orbit and of the access to it, are only indirectly addressed by the Outer Space Treaty, as it does not define "access" or the "geostationary orbit." The fundamental problem in extending the rule of law into a new environment, is how to coax its legal framework to encase ill-defined policy spaces.

Access to the geostationary orbit is addressed by three principles contained in Articles I and II of the Outer Space Treaty. The first principle is the "Freedom of Exploration and Use" stated in Article 1, paragraph 2:

> Outer space ... shall be free for exploration and use by all States without descrimination of any kind, on a basis of equality and in accordance with international law, ...

The second principle, stated in Article 2, is the prohibition of national appropriation or extension of sovereignty into outer space:

> Outer space is not subject to national appropriation by claim of sovereignty, by means of use or occupation or by any other means.

The third principle establishes the "ends" or goals of outer space exploration and use:

> The exploration and use of outer space ... shall be carried out for the benefit and in the interest of all countries, irrespective of their degree of economic or scientific development, and shall be the province of mankind.

1. Open Access

"Free and open exploration and use" as stipulated by the Outer Space Treaty mandates "open access" to the geostationary orbit. The term "use,"

> ... in the legal sense refers to the enjoyment of property which usually results from the occupancy, employment, or exercise of such property. Usually also there is an element of profit, benefit, or some other measure of advantage accompanying the use. [3]

The definition of "use" is pivotal to answering the question: "Open access to whom?" While the Outer Space Treaty asserts "free exploration and use" to all *states*, the Treaty does not mention non-governmental or private enterprise entities and organizations. Are they excluded from legally-sanctioned outer space exploration and use?

The issue of "open access to whom" surfaced in the early drafting stages of the Outer Space Treaty. The Soviet Union proposed that only states be allowed to operate in outer space. A draft submitted by the Soviet Union in 1962 stipulated that:

> All activities of any kind pertaining to the exploration and use of space shall be carried out solely and exclusively by States ... [4]

The Soviet position was later explained by delegate G.P. Zhukov, who wrote in 1974:

> ... the USSR's position on this matter was dictated by its "justified fear" that granting freedom of arms in space to Western private industry would encourage "the kind of activity correctly characterized as piracy." [5]

The Soviet view identifies a major loophole in the legal fabric of the emerging legal regime. While states which are parties to the Treaty agree to conduct their activities according to their own domestic and international law, non-governmental entities are not automatically covered by the same legal umbrella when operating in the legally undefined geographical region — outer space. This question is further blurred by the Treaty's lack of a legal demarcation of outer space; the point where national airspace sovereignty ends and the legal commons of outer space begins. In cases of interference or conflict arising from activities performed by states and non-governmental entities, is one party negotiating under one set of laws, namely international law, and the other, assuming that states have not adopted municipal law to implement international agreements, under none?

This could be particularly troublesome in cases involving communications satellites in the geostationary orbit. In situations where one state's satellite and a satellite operated by a foreign private firm cannot function due to mutual interference, must the state coordinate and negotiate with foreign private individuals who may have no legal obligations or responsibility to do so? One deep-seated concern was that private broadcasting entities might use high-powered satellites to beam programming into countries without the affected state's approval. Unless the rule of law could be extended so as to include non-governmental entities operating in outer space, ITU procedures for registration and coordination of communications satellites used by private firms would have no legal basis.

In contrast to the Soviet position, the United States viewed outer space as an area for private enterprise to develop and use. This position was set into law and official United States policy by the passage of the Communications Satellite Act of 1962 [6]. The Act established the Communications Satellite Corporation (called COMSAT), a quasi public-private corporation which would construct and operate a satellite communications system. The intent of Congress was to provide for the "widest possible participation by private enterprise" in a "communications satellite corporation for profit." [7]

In the "private enterprise" view, the inclusion of the term "use" in Article 1 and its meaning (defined above)does allow private entities to operate for a profit in outer space. Furthermore, the reference to "all States" in the same paragraph is seen by legal scholars not to "preclude the exercise of this circumscribed freedom by entities other than States." If that had been the intent, they could have inserted the word "only" to make the phrase read "only by States." [8]

The Soviet Union and other countries predisposed against non-governmental entities operating in outer space did not prevent the Committee on Peaceful Uses of Outer Space (COPUOS) from arriving at a consensus on the Outer Space Treaty draft. The compromise adopted established significant barriers to unrestricted exploration and use by non-governmental entities, i.e., private enterprise. As one legal scholar commented:

> Private enterprise, as subject of international law, will have to be strictly regulated by their respective States since the States bear international responsibility for national activities in outer space under the Treaty. This burden of state responsibility is made heavier by a duty to register any space object launched from its territory. [9]

Responsibility, liability, and registration connect the actions of private enterprise in outer space to states under the Treaty. Articles 6 through 8 extend the rule of law to private non-governmental and intergovernmental entities operating in outer space. Article 6 of the Outer Space Treaty stipulates that:

> States Parties to the Treaty shall bear international responsibility for national activities in outer space, including the moon and other celestial bodies, whether such activities are carried on by governmental agencies or by non-governmental entities, and for assuring that national activities are carried out in conformity with the provisions set forth in the present Treaty. The activities of non-governmental entities in outer space ... shall require authorization and continuing supervision by the appropriate State Party to the Treaty. When activities are carried on in outer space ... by an international organization, responsibility for compliance with this Treaty shall be borne by the States Parties to the Treaty participating in such organization.

In addition, Article 7 provides that:

> Each State Party to the Treaty that launches or procures the launching of an object into outer space ... and each State Party from whose territory or facility an object is launched, is internationally liable for damage to another State Party to the Treaty or to its natural or juridical persons by such object or its component parts on the Earth, in air space or in outer space ... [10]

The rule of law is extended to each object launched by the provisions of Article 8, which stipulates:

> A State Party to the Treaty on whose registry an object launched into

> outer space is carried shall retain jurisdiction and control over such object, and over any personnel thereof ... Ownership of objects launched into outer space ... and of their component parts, is not affected by their presence in outer space ... [11]

As the quoted Articles demonstrate, private companies are granted access to the space environment under restricted conditions. First, "authorization and continuing supervision" of the non-governmental entity by a state is required. Secondly, the sponsoring state assumes responsibility and liability for activities by the private firm. The activities must conform to the state's treaty obligations as stipulated by the Outer Space Treaty. In addition, the sponsoring state is liable for any damages incurred as the result of such activities. This aspect was reinforced by the 1972 Liability Treaty. Under the provisions of the Treaty, "a rule of absolute liability is imposed upon the launching state for damages causes by the space object." [12]

Article 9, while not imposing restrictions on the activities of private organizations in space, does clearly stipulate that states bear the responsibility for the actions of entities operating under its sponsorship and jurisdiction. In cases where activities of private entities and states come into conflict, it is the sponsoring state which must consult with the other states, not the non-governmental organization. Non-governmental entities and organizations do not occupy the same legal status of states in outer space. In the final legal sense, all activities in space are performed under the jurisdiction of states [13].

In sum, "free exploration and use" can be translated as "open access" to the geostationary orbit. The stated opinions of many legal scholars and past practice indicate that open access applies to both governmental and non-governmental entities.

Private enterprise is allowed by the Outer Space Treaty to operate in the geostationary orbit and extract a profit under the concept of "use" provided for by the Treaty. However, non-governmental (and, presumably international intergovernmental) entities using the geostationary orbit do so under the legal sponsorship and supervision of national governments.

In so doing, states assume responsibility and liability for the activities performed. In cases of interference or conflict, states negotiate directly with other states, as private entities are considered legally subordinate to the sponsoring state. Furthermore, since all objects in the geostationary orbit have to be launched to get there, the provisions of Articles 7 and 8 regarding registration and consultation firmly extend the rule of law to include non-governmental entities, operations, and objects in space. Jurisdiction and liability are established by the act and location of launching.

2. Non-Appropriation of the Orbit/Spectrum Resource

Article 2 prohibits the appropriation of outer space. The "open access" provisions of Article 1 require the ban on appropriation in Article 2 for access to remain "open" to all states. Free and open exploration as well as the use of outer space is possible only if such activities are not pre-empted by other users. This is precisely the intent of Article 2. Open access necessitates non-appropriability. While the claim of sovereignty is an obvious action, non-appropriation "by means of use or occupation, or by any other means" is not. As noted by legal scholars,

> The ability of States to use, occupy, and pre-empt an area without claims to sovereignty has ... been repeatedly demonstrated in history. [14]

2.1. To Whom Does the Ban on Appropriation Apply?

In the preceding section, "open access" is not found to be limited exclusively to states. Inter-governmental organizations, such as INTELSAT, and private firms, such as Western Union International, can operate and use outer space as long as it is done under the supervision and sponsorship of states, who assume liability and responsibility for their activities under international law.

However, the ban on appropriation in Article 2 is explicitly worded against *national* appropriation. The explicitness of the ban against national appropriation is seen by some legal scholars as posing the potential loophole that entities other than nations could appropriate outer space. An example of the more restricted interpretation of Article 2 is shown by the view taken by Professor Gorove in 1977:

> [T]he Treaty in its present form appears to contain no prohibition regarding individual appropriation or acquisition by a private association or an international organization, even if other than the United Nations. Thus, at present, an individual acting on his own behalf or on the behalf of another individual or a private association or an international organization could lawfully appropriate any part of outer space ... [15]

Gorove's view is quite consistent with interpretations of Article 1 and "open access." Since access was not denied to entities other than states, then it must be allowed. In the same way, since appropriation is not explicitly prohibited to non-governmental entities, the omission may mean appropriation by them is neither banned nor allowed.

However, just as "open access" was extended within a regime including non-governmental entities under restricted conditions, might the ban on appropriation also follow the path taken by "open access" to non-governmental entities? Professor Carl Christol takes an expansionary view on Article 2. Christol interprets the ban on appropriation as creating a *res communis* regime. Under a *res communis* regime, all of outer space is already "owned" by all states, hence no one (state, private organization, or intergovernmental organization) can claim or appropriate any portion of it. As Christol writes:

> Article 2 is based on the *res communis* principle. From this it follows that under the existing legal regime no juridical or natural person can establish sovereign or proprietary rights in either the space environment or in its resources.

At this point, Christol expands the ban on appropriation to intergovernmental organizations, such as the ITU.

> If states are prevented from asserting such sovereign or proprietary rights for themselves, if follows that states may not claim such rights in the exercise of sovereign or proprietary rights on behalf of others, such as an international intergovernmental organization [i.e., the ITU].

In other words, states making up the ITU may not use the organization as a vehicle for appropriating outer space, and in particular, the geostationary orbit.

> This conclusion is strengthened by accepting the view that the "by any other means" provision of Article 2 constrains states, and their nationals — who would have no greater rights than states — as well as international intergovernmental organizations. The latter would have to rely on states, rather than on their own independent legal authority as a legal person, to exercise quasi-sovereign authority. They could not assert such authority on their own account. [16]

To pursue Christol's arguments further, an international, intergovernmental organization which on its own authority allocates slots in the geostationary orbit, "represents [in] itself an 'appropriation'."

> Commitment to the prohibition in Article 2 against acquisition of sovereignty by any other means would prevent the conversion of an *a priori* grant to any recipient through the ITU system into a basis for asserting national sovereignty respecting the allotted frequency or orbital position. [17]

In sum, Christol extends the ban on appropriation to all entities having access to outer space and the orbit/spectrum resource. States sponsoring non-governmental private organizations must supervise the activities so that such use does not violate international law, including activities constituting appropriation. States also may not transfer their sovereign prerogatives to an intergovernmental organization, granting it the authority to appropriate and to allocate portions of outer space (i.e., the geostationary orbit) back to the states which then could treat such allocations as sovereign "territory." This interpretation would seem to rule out *a priori* allotments of orbital bandwidth as envisioned under some detailed planning schemes. It also complicates the legal status of common-user satellite organizations, such as INTELSAT or GLODOM arrangements, where it remains to be settled whether portions of the resources can be allocated as intergovernmental organizations, or whether only allotments assigned to them by member states can be used [18]. In sum, it still remains to be determined under what conditions "use," as guaranteed by Article 1, constitutes appropriation as prohibited by Article 2.

2.2. Occupation and Use as Appropriation

Occupation and use may become appropriation if there is a sense of permanence or ann intent to appropriate or claim sovereignty over an area of outer space.

As defined by Christol, "appropriation"

> . . . is used most frequently to denote the taking of property for one's own or exclusive use with a sense of permanence. [A]ny use involving consumption or taking with intention of keeping for one's own exclusive use would amount to appropriation. [19]

Let us now explore the concept of "time" in terms of occupation and use and how it applies to appropriation.

Geostationary satellites currently have "engineering lifetimes" averaging about seven years before station-keeping fuel is expended. After that time, ground controllers are unable to maintain the satellite's position in the geostationary orbit or to accurately keep the satellite's antennas pointed at their intended service areas. At the present time, technological constraints do not allow a permanent functioning presence in a particular orbital slot. The satellite is only temporarily occupying the slot, a fact recognized by the ITU Radio Regulations. Once a satellite is no longer operational, its registration of orbital location and frequencies with the IFRB lapses. In addition, that user has no automatic prerogative or right to continue using that particular slot without going through IFRB notification and coordination procedures.

New technology may change this situation. Programs undertaken by both the Soviet Union and the United States are intended to test techniques for refueling space vehicles. The Soviet Union has proven through the Salyut series that they can periodically launch space "tankers" that dock with the Salyut space station, transferring supplies and fuel. The operational lifetime of a Salyut station is, in effect, unlimited. The United States Space Transportation System (Space Shuttle), has similar mission objectives and capabilities. On Shuttle mission 10, launched in February 1984, astronauts left the craft as a test to see how repairs or refueling might actually be accomplished. This technology can increase a satellite's operational lifetime making it feasible to maintain a permanently manned or unmanned station in low earth orbit, and eventually, in the geostationary orbit.

2.3. How Long Is Appropriation?

Does a geostationary satellite with an unlimited operational lifetime represent an appropriation of outer space and the geostationary orbit? The issue is no more clearly posed than with the proposals for "Geostationary Orbital Platforms," or "Geoplats." [20] Geoplats are large space stations designed to fulfill several different functions from the same orbital slot. In addition to the communications function, a geoplat could also serve as a weather reconaissance station, a military nuclear detection function, or an intersatellite relaying station. Geoplats would use a modular construction scheme on a common bus; that is, each function would be served by a plug-in module, all of which would share a common power supply and station-keeping systems. In this way, economies of scale could be realized. The development of a space "tug" is intended to permit the automatic refueling, and possibly, the repair or modification of the geoplat during its essentially unlimited lifetime. A geoplat would be a permanent presence in the geostationary orbit.

If appropriation is an act of presence that pre-empts use or occupation of a commons resource by others, a geoplat would represent appropriation only when it pre-empts other users and when there is an intent to appropriate. A permanent presence by itself is not viewed as necessarily constituting appropriation [21].

As stated above, use or occupation represents appropriation when such activities pre-empt others from gaining access to outer space. This is particularly poignant regarding a favorable and crowded area of outer space —the geostationary orbit. As Smith points out,

> Generally, space objects are constantly in motion and thus do not "appropriate" particular areas of outer space in violation of the Article 2 prohibition. However ... objects placed in geostationary orbit maintain the same location relative to the earth's surface. In theory, therefore, it could be argued that space stations "appropriate" segments of the geostationary orbit. It is in this context that the meaning of Article 2 can greatly impact the ownership and operation of space stations that utilize the geostationary orbit. [22]

However, the presence of a space station in a particular slot only becomes problematical once it is perceived that such occupation pre-empts others. "The difficulty begins the moment when the orbit is filled to capacity — in this case national appropriation would occur by means of use or occupation." [23]

2.4. Conditions of Resource Scarcity or Saturation

If the geostationary orbit is filled to its electromagnetic capacity, additional users cannot gain access to an orbital slot without causing interference (electromagnetic) to other users. The use of slots in saturated segments could prevent others from using or gaining access to the resource without harmful interference. In other words, does use of limited and particularly favorable areas of outer space such as the geostationary orbital slots, under conditions of saturation where occupation by one may pre-empt use by another, constitute appropriation? Legal interpretations of the Article 2 ban on appropriation must perform a balancing act between "open access" as provided by Article 1, and the conditions, such as time and resource saturation, under which use can be considered "appropriation."

In this weighing of use against appropriation, legal scholars attach great weight to the intent of the user and the end result of such use. If there is no evidence of an intent to appropriate or to extend sovereignty, and if such use benefits mankind, then the occupation and use of geostationary slots which may pre-empt their use by others, does not in itself represent appropriation. As Christol writes:

> If, as suggested, a primary purpose of the Principles Treaty was to allow states to enjoy the peaceful use of the space environment, then the *intent* of a using State becomes important [emphasis added]. [24]

2.5. *Intent to Appropriate*

States may claim, occupy, and use orbital slots for various reasons. They may wish to extend their national sovereignty to segments of the orbital arc as was proposed in the Bogota Declaration. States may exploit their technological capabilities and the present ITU allocation, registration, and coordination procedures by claiming slots and frequencies in advance, thereby pre-empting their use by other nations for years to come [25].

States may utilize their technology or combine technologies as with geoplats so that use and occupation are non-exclusive. Several nations could plug in their different modules into a geoplat, making possible a multinational presence in a particular orbital slot.

As Christol writes:

> If the intent to assert an exclusive right to a given use, and if that use is designed to and is carried out, as provided in Article 1 of the Principles Treaty, "for the benefit and in the interests of all countries ... then such conduct would not consist in an appropriation by the using State. [26]

3. *Province of Mankind*

Outer space is the "province of mankind." This statement in the Outer Space Treaty [27] represents compromise wording reached during the drafting stages between LDCs and HDCs in the Committee on Peaceful Uses of Outer Space [28].

While the LDCs desired specific reference to the goal of space exploration and use "for the benefit and in the interests of all countries, irrespective of their degree of economic or scientific development;" the HDCs, on the other hand, desired greater vagueness embodied in the phrase, "province of mankind." The "vagueness" desired in the HDC wording reflects the diversity between national and private space programs for differing goals. The phrase is credited with giving the Treaty greater "cohesion" in narrowing what would otherwise be a rift in the stated goals of access to the space environment. Consensus was achieved.

Challenge to the Commons: The Bogota Declaration

In 1976, delegations from eight equatorial countries — Brazil, Colombia, Congo, Ecuador, Indonesia, Kenya, Uganda, Zaire — meeting in the Colombian capital city signed the so-called "Bogota Declaration." Regardless of the document's disputed legal merits, it has been successful in focusing attention to the orbit's legal status. With the Bogota Declaration, the issue of access to the orbit became even more strongly identified with the North-South dialogue as encompassed within the New World Economic Order issue arenas.

In the document, the equatorial countries claim national sovereignty over the segments of the geostationary orbit "corresponding" to their "national terrestrial, sea, and insular territory." According to the document's interpretation of the right of access to the geostationary orbit:

> Devices to be placed permanently on the segment of a geostationary orbit of an equatorial state shall require previous and expressed authorization on the part of the concerned state, and the operation of the device should conform with the national law of that territorial country over which it is placed. [29]

While the Declaration's signers purport that extention of their sovereign rights to the geostationary orbit is "directed towards rendering tangible benefits to their respective people," its real aim, observers note, is to establish a legal and political basis with which to charge "rent" for use of the resource. with this deft political maneuver, the inferred goal of the Declaration fits neatly onto NIEO objectives calling for a redistribution of global wealth in general, and of the benefits and profits of satellites' exploitation of the geostationary orbit, which, under existing rules is "used to the greater benefit of the most developed countries," in particular.

Here, the Declaration refers back to the "interest and benefit of all states" and "province of mankind" provisions in Article 1 of the Treaty. Using this as a legal basis, politically, the Bogota Declaration is an "exceptionally successful tactic for use as a bargaining chip in geostationary orbit access negotiations in the ITU and COPUOS." [30]

Customary International Space Law

In a balance of power perspective, international law mirrors the configurations of technological capabilities and political, economic, and strategic interests of countries. Customary and substantive international law is created and established through the actions (or inactions) of sovereign states. International law exists because nations find compliance in their national interest, or see no other acceptable or less-costly policy alternative to cooperation within the rules established in an international legal regime. In other words, the rules will be followed to the extent that they correspond to the distribution of power and each nation's perceived self-interest.

The Treaty and customary international law establish that outer space is a region outside nations' jurisdictional boundaries. Will this legal status continue? At the beginning of the Space Age, countries did not possess the means to enforce or extend their sovereign prerogatives into the unexplored environment. It was at least tacitly acknowledged by the space powers that outer space was a region outside of a nation's airspace and territorial sovereignty. Neither the Soviet Union nor the United States protested flights over their territory by the other nation's satellites. In doing so, the space powers established a legal precedent and customary international law recognizing outer space as a distinct legal region, analogous to the high seas. Outer space is a global commons.

However, this could change. The development of anti-satellite weapons now makes it possible to destroy satellites from the earth or from space battlestations. The questions becomes: will customary and substantive sources of international law be able to preserve outer space as a legal, global commons as nations acquire the means to extend their sovereign control beyond national airspace?

In this respect, the United States' reluctance to establish a boundary between sovereign airspace and outer space may reduce the legal protection afforded to satellites that would orbit within the region of the global space regime. In a more "Machiavellian" sense, it leaves the option of being able to shoot down satellites without incurring the cost of violating the Treatly open. The lack of a formal boundary may create uncertainties, leaving satellites in a sort of "legal limbo" regarding their legal status and protection.

In sum, the Outer Space Treaty establishes a right of open and free access to the geostationary orbit for all nations. The Treaty and customary international law support a finding that the geostationary orbit lies outside the jurisdictional boundaries of states in outer space, although the lower limit of outer space is not formally defined or delimited. There is, however, no explicit legal basis to support a contention that the Treaty stipulates "equitable access" to the geostationary orbit. The "province of mankind" and the ban on appropriation provisions indicate the intent of the Treaty's authors and ratifying states — that is, the benefits stemming from use of the outer space environment, including the geostationary orbit, are to accrue to all states, although the form of such benefits, e.g., access, are not explicitly stated.

REFERENCES

1. S. Neil Hosenball, "Who Owns What? Earth Law for Space," *IEEE Spectrum*, September 1983, p. 76.
2. Treaty on the Principles Governing the Activities of States in the Exploration and Use of Outer Space, Including the Moon and other Celestial Bodies (Outer Space Treaty). Open for signature January 27, 1967, 18 U.S.T. 2410, T.I.A.S. 6347, 610 U.N.T.S. 205, entered into force for the United States on October 10, 1967.
3. Stephen Gorove, *Studies in Space Law: Its Challenges and Prospects*, (Leyden: Sijthoff, 1977), p. 54.
4. UN Document A/5181 Annex III A (1962), cited in Arthur Dula, "Free Enterprise and the Proposed Moon Treaty," *Houston Journal of International Law* 2:3 (1979), p. 5.
5. *Ibid.*
6. Communications Satellite Act of 1962, 47 U.S.C. 704-44.
7. Arthur Dula, "Regulation of Private Commercial Space Activities," in *The 24th Colloquium on the International Law of Outer Space*, Rome, September 1981, p. 32.
8. Gorove, p. 50.
9. Gijsbertha Reijnen, "Outer Space Law and Private Enterprise in Outer Space: An International Perspective," in *Houston Journal of International Law*, 2:15 (1979), p. 69.
10. Article 7, Outer Space Treaty.
11. Article 8, Outer Space Treaty.
12. Martin Menter, "Commercial Participation in Space Activities," *Journal of Space Law*, 9:1 (Fall 1981), p. 56.
13. Article 9 of the Outer Space Treaty indicates that:

    If a State Party to the Treaty has reason to believe that an activity or experiment planned by it or its nationals in outer space ... would cause potentially harmful interference with activities of other States Parties in the peaceful exploration and use of outer space ... it shall undertake appropriate international consultations before proceeding with any such activity or experiment.

    Article 13 of the Outer Space Treaty prescribes rights and duties of international intergovernmental organizations:

    The provisions of this Treaty shall apply to the activities of States Parties to the Treaty in the exploration and use of outer space ... whether such activities are carried on by a single State Party to the Treaty or jointly with other States, including cases where they are carried on within the framework of international intergovernmental organizations.

Any practical questions arising in connection with activities carried on by international intergovernmental organizations ... shall be resolved by the States Parties to the Treaty either with the appropriate international organization or with one or more States members of that international organization, which are Parties to this Treaty.

14. Houston S. Lay and H.J. Taubenfeld, *The Law Relating to the Activities of Man in Outer Space* (1970), p. 78.
15. Gorove, 1977, p. 81.
16. Carl Q. Christol, "National Claims for Using/Sharing of the Orbit/Spectrum Resource," in *25th Colloquium on the International Law of Outer Space*, Paris, 1982, p. 9, fn. 35. He has also found support for this view in the terms of Article 13 of the 1967 Treaty.
17. *Ibid.*, p. 7.
18. See Wilson P. Dizard, "Space WARC and the Role of International Satellite Networks," paper presented at the Center for the Strategic and International Studies, Georgetown University, August 1984.
19. Christol, 1982, p. 82.
20. See Delbert D. Smith, *Space Stations: International Law and Policy*, (Boulder: Westview Press, 1980), and Delbert D. Smith and Martin A. Rothblatt, "Geostationary Platforms: Legal Estates in Space," in *Journal of Space Law*, 10:1(Spring 1982):31-39; and Delbert D. Smith, "International Utilization and Management of Space Systems," in *Houston Journal of International Law*, 2:1(1979):113-129.
21. Arthur Dula, "Regulation of Private Commercial Space Activities," *24th Colloquium on the Law of Outer Space*, Rome, September 1981, p. 32. Dula writes:

    Keeping a satellite in orbit for an extended period, e.g., 30 years, might, however, indicate a "sense of permanence" constituting an act of appropriation. The focus then must switch to the intent of the launching authority. Article 1 of the Outer Space Treaty states that "outer space shall be free for exploration and use ... " Such use, the United States has pointed out, might be prolonged without constituting an express intent to perpetuate a State's presence or authority.

22. Smith, *Space Stations*, p. 103.
23. Oscar Fernandez-Brital, "Geostationary Orbit," *21st Colloquium on the Law of Outer Space*, (1979), p. 14.
24. Christol, 1982, p. 45.
25. See David S. Myers, "Common Interest' and 'Non-Appropriation' in

Outer Space: Political Interpretation of Legal Principles," in *International Relations* (London) 6:3 (May 1979):539. Myers writes:

> They may be able to pre-empt areas or resources effectively to the exclusion of others. In this situation, the State may be in a position to gain considerable economic or strategic advantage and at the same time reap political benefit by not claiming sovereignty over the area.
>
> This may particularly be the case with the present ITU procedures based on the first-come first-served principle. Advanced technology states may initiate coordination and registration procedures for saturated sections of the arc thereby pre-empting the use of orbital slots by other nations for perhaps decades with long-life spacecraft. These actions are not taken in the name of national sovereignty, but rather under the innocuous rubric of ITU Radio Regulations.

26. Carl Q. Christol, "The Geostationary Orbital Position as a Natural Resource of the Space Environment," *Netherlands International Law Review* 26:1(1979):11.
27. Article 1, para. 1:

    > The exploration and use of outer space, including the moon and other celestial bodies, shall be carried out for the benefit and in the interests of all countries, irrespective of their degree of economic or scientific development, and shall be the province of mankind.

28. For a more detailed account, see Christol, 1982, p. 45; and Gorove 1977, p. 56.
29. Bogota Declaration, Article 2, sub. D.
30. Comments by Amanda Lee Moore, international space lawyer, during a Legal Symposium at the Non-Governmental Organizations Conference at UNISPACE '82, 18 August 1982, Vienna.

CHAPTER 4

THE INTERNATIONAL TELECOMMUNICATION UNION: WHERE TECHNOLOGY GETS DOWN TO THE BUSINESS OF POLITICS

> Perhaps of greater importance is the distributive function of the ITU. It provides access for all countries to the radio frequency spectrum and the geostationary orbit. It also ensures the distribution of the benefits from the use of the spectrum and the orbit among member countries. Even though the ITU carries out these tasks at a "technical" level the truth is these are political activities. *The distribution of any resource is eminently a political act* [emphasis added]. [1]
>
> - *Armando Vargas*

Introduction

The International Telecommunication Union (ITU) is the institutional core of an emerging global regime for the orbit/spectrum resource. The ITU, which traces its origins to the founding of the International Telegraph Union in Paris in 1865, is the oldest continuously operating intergovernmental organization. The ITU's longevity and success can be traced to the universal need to communicate, which provides compelling incentives for states to join and participate in the organization.

Starting with the telegraph, and continuing with the telephone and other "wire" services, the organization has promulgated the necessary technical standards that make it possible for telecommunications equipment of one country to interconnect with all others. The advent of "wireless" telecommunications at the end of the century required an even greater degree of standardization due to its use of a "commons" resource — the radio spectrum. Through this path, the ITU (actually its predecessor organization — the International Radiotelegraph Union, or IRU) assumed the task in the early 1900s of standardizing the use and allocations of the radio spectrum resource.

Space exploration extended telecommunications into outer space, a development which expanded the ITU's spectrum jurisdiction to the geostationary orbit and beyond, as communications satellites, radio astronomers, planetary probes, manned moon missions, and other space-based services needed and demanded spectrum allocations and protection from interference.

The need for spectrum management has existed since the spark-gap days of radio when more than two operators in the same region would cause each other communications-disrupting interference. As transmitters grew more powerful, permitting intercontinental radio services at the turn of the century, harmful interference also became a matter of international concern.

Since 1903, the radio spectrum has been managed through multilateral agreements between user states or their telecommunications entities. The first international conference was held in Berlin in 1903. This "Preliminary Conference Concerning Wireless Telegraphy" produced guidelines stipulating that stations should operate without causing interference to other stations. Three years later, delegations from 29 countries convened once again in Berlin where they promulgated the International Radiotelegraph Convention and annexed regulations, establishing the International Radiotelegraph Union.

The Berlin Conference allocated portions of the spectrum resource to specific radio "services" for the first time. For example, frequencies below 188 kHz were set aside for long distance communications, while frequencies between 188 and 500 kHz were reserved for government services. International Radiotelegraph Conferences were later held in London (1912) and in Washington, D.C. (1927).

Increased awareness of the need for cooperation and standards in radio applications resulted in the formation of the International Technical Consultative Committee for Radioelectric Communications at the Washington Conference, which was the forefunner of the present CCIR.

An innovation of the 1927 Conference was the Table of Allocations. The table listed the spectrum allocations to the various radio services. It also specified whether frequencies could be used by a particular radio service in regional or worldwide applications. Most importantly, the Table of Allocations stipulated "through which mechanisms a nation could acquire a *right* [emphasis added] for its radio stations to use specific radio channels without harmful interference from the stations of another nation." The table later became the Master Frequency Register, administered by the International Frequency Registration Board (IFRB). At these meetings and conferences, the participating administrations attempted to resolve the question of what "kind of administrative scheme . . . should be adopted for vesting spectrum user rights in frequency usage." It is the issue challenging the ITU today, for it constitutes political access to the spectrum resource.

In 1932, the administrations participating in the 13th International Telegraph Convention and 4th International Radiotelegraph Convention agreed to merge their organizations under the overarching structure of a new International Telecommunication Union. The Madrid Conferences promulgated the International Telecommunication Convention, the document which serves as a written constitution for the organization.

The postwar Atlantic City Conferences in 1947 adopted changes that have shaped the structure of the organization up to the present. The Plenipotentiary Conference produced a new International Telecommunication Convention which integrated the ITU into the United Nations system as a specialized agency. This change required replacing the Swiss-run bureau in Bern with the establishment of an international secretariat in Geneva. Most importantly, the IFRB was formed and charged with the responsibility of over seeing the process of registering and coordinating use of the radio spectrum.

In addition, the successor to the 1927 Radio Committee, now called the International Radio Consultative Committee (abbreviated as the "CCIR" corresponding to its title in French) became a permanent body within the ITU. Its counterpart for "wire" telecommunications services, the International Telegraph and Telephone Consultative Committee, or CCITT, was also made a permanent organ within the ITU [2].

Although a multitude of intergovernmental entities are now involved in various aspects of space telecommunications, the ITU is the organization directly responsible for allocating, regulating, and managing the orbit/spectrum resource and is therefore in the middle of the political debate over satellite communications. The ITU 1985/88 World Administrative Radio Conferences (Space WARC or ORB-85/88) will determine how the orbit/spectrum resource will be used well into the 21st century.

The ITU legislates political access to the orbit/spectrum resource through its complex "federal" structure, involving both legislative and juridical processes and entities. "Political access," in sum, is the internationally recognized *right* to use a portion of the orbit/spectrum resource, free from harmful interference. In reality, the ITU's policy-making process is a system for allocating and administering "rights" to the orbit/spectrum resource.

In a three-step analysis, this chapter will look at both the institutions and the policy-making process involved in determining political access to the orbit/spectrum resource. First, the outlines of the controversy and debate over the orbit/spectrum resource, as viewed from the vantage point of the ITU, will be drawn. Next, the policy-making structure and organs of the ITU will be analyzed as they affect the Space WARC decision-making environment. Thirdly, the ITU's policy history regarding the issue of access to the orbit/spectrum resource will be reviewed in order to indentify salient trends that may indicate likely schemes for managing the resources.

Technology, Politics, and States' Interests

The ITU is where politics meets technology head-on, producing an inevitable policy tension between the dictates of technology and the motives of states. The organization's policy output, designated the "Radio Regulations," are derived from immutable physical laws of electromagnetic radiation and the constantly evolving practices of radiocommunications engineering. These indicate what is technically feasible by taking into account how telecommunications systems interact and interfere with each other. The goal of the Radio Regulations is to promote "efficient" use of the orbit/spectrum resource.

However, beneath their antiseptic technical veneer, the Radio Regulations have conspicuous political consequences as they determine which users shall enjoy the benefits of legally recognized and protected access to the orbit/spectrum resource. The onrush of power-enhancing satellite communications technologies have intensified the tension between states' compulsion to maximize sovereign prerogatives regarding orbit/spectrum resource access and use, and their concurrent need to cooperate in orbit/spectrum resource management arrangments — an unresolved strain politicizing the ITU's procedures for allocating access rights to the orbit/spectrum resource.

The Controversy over Access in the ITU

Today the ITU is an institution in transition. Its long-standing organizational and policy focus on esoteric technical standards and frequency allocation procedures, intended to promote the most efficient use of the orbit/spectrum resource, faces mounting challenges and demands for reform from an LDC majority. This majority is demanding new rules and procedures designed to ensure all countries, regardless of their level of technical capability, "equitable access in practice" to the orbit/spectrum resource and satellite communications.

On the institutional level, there is a political battle over access rights and the organization's procedures for administering those rights. Under the current Radio Regulations, the right to use portions of the spectrum or orbit resources with legal protection from harmful interference is granted by ITU procedures to users on an *a posteriori* basis. This is the trademark of the so-called "first-come, first-served" regime. The first user of a frequency or orbital slot enjoys priority rights to that frequency or orbital slot in any subsequent IFRB-administered "coordination" procedures instigated by latecomers.

Coordination procedures are prescribed by the Radio Regulations for cases where an additional satellite system introduces interference over a predetermined "trigger" value to already operating systems. In practice, the onus and expense is clearly placed upon the latecomer to adapt or change his space and earth segment system parameters to fit the interference environment created by earlier users. As more and more satellite systems fill up portions of the arc, coordination has become a greater burden as resource congestion imposes higher technical standards and costs.

The LDCs, as latecomers to telecommunications technology, fear that their future plans for regional or national satellite systems are endangered by an increasingly crowded geostationary orbit that is being filled by HDC satellite systems. In the LDC scenario, by the time they have the means or the need to establish their own satellite systems, the entry price to a congested orbit/spectrum resource will be higher than they can afford. The developing countries propose to reserve sections of the orbit/spectrum resource on an *a priori* basis for their eventual use. In this way, they believe, entry costs can be frozen by arbitrarily freezing the level of technology necessary to achieve interference-free access.

The LDCs propose to abolish "first-come, first-served" in favor of detailed *a priori* planning. Under such an arrangement each country would receive, on the basis of its membership in the ITU, an allotment of orbital slots and frequency channels. Only in this way, the LDCs argue, can "equitable access" be "guaranteed in practice" to the increasingly crowded orbit/spectrum resource.

The HDCs counter such arguments by pointing to prior experience which shows that no satellite system has ever been denied ITU-approved access. In addition, the HDCs point to the constantly evolving technological innovations that continue to expand the communications capacity of the satellite systems utilizing the orbit/spectrum resource. Costs of satellite circuits are declining as the number of circuits handled by one satellite increases. This trend might reverse itself, HDC experts contend, if a rigid, detailed plan were put into effect, artificially restricting the usable amounts of the resource, and thereby driving up the access price as supply declined.

The access controversy is a jigsaw puzzle. Moreover, it is one of extraordinary complexity due to its rapidly changing pieces that resist easy or permanent arrangments. Are dynamically evolving technologies making the orbit/spectrum resource more or less accessible to LDCs? Is access polticizing access, or are there other issues and motives that overpower the technical aspects?

Some observers of the orbit/spectrum resource access debate contend that the latter is true. They see the LDC proposals for reforming access-granting procedures as part of an overall strategy to convert the ITU from a technically-oriented and relatively passive standards-setting body to an aggressive international development agency which uses its budget to finance satellite and other telecommunications projects in the Third World.

From the HDC perspective, LDCs are using the orbit/spectrum access issue as a power fulcrum from which to gain leverage in negotiations on other space-related matters in the ITU. In other words, the LDCs are using access to influence the use dimension of space and telecommunications — areas where they have less direct influence due to their technological powerlessness. The ITU, as regulator of the geostationary orbit, is a favorable forum since the LDCs constitute a voting majority. However, such negotiating maneuvers by themselves have little power to influence policy, were they not predicated on very real technical and economic vulnerabilities of HDCs. The issues of free flow of business information over national boundaries and cooperative use of the spectrum resource are factors forcing the HDCs to listen to LDC demands in the ITU. In sum, we can see how use is politicizing access and *vice versa.*

Organizational Structure of the ITU

The ITU is unique among international organizations in the United Nations system as an institution without a permanent charter or constitution. Instead, at approximately seven-year intervals, Plenipotentiary Conferences are convened with the authority to revise or rewrite its supreme document, the ITU Convention, or even to disband the organization. Since World War II, Plenipotentiary Conferences have been held in 1947, 1952, 1959, 1965, 1973, and, in 1982, in Nairobi, Kenya, the first such conference held in a Third World country [3]. The organizational success and longevity of the ITU may be attributed to its structural flexibility, a characateristic which allows the organization to change with the technology it regulates.

The ITU Convention, once ratified by a state, is recognized as international law and has the same binding authority. The Convention establishes the administrative hierarchy, jurisdictions, duties, funding, and responsiblilities for the various entities composing the ITU. The Convention is the legal glue binding the ITU's federal structure of semi-autonomous bodies together.

The ITU's permanent organs under the Plenipotentiary Conference and Convention are the General Secretariat and the Administrative Council. These form the backbone of the bureaucratic machinery concerned with the day-to-day operations of the ITU. The actual policy decisions concerning telecommunications are made by the International Frequency Registration Board, the International Radio Consultative Committee, the Internaional Telephone and Telegraph Consultative Committee, and the Administrative Conferences for either radiocommunications, called World Administrative Radio Conferences, or for wire-based telecommunications, called World Administrative Telephone and Telegraph Conferences.

As shown in Table 4.1, the central policy-making body is the Administrative Radio Conference, called either a WARC, for World Administrative Radio Conference, or RARC, for Regional Administrative Radio Conference, depending on the global regions involved [4]. WARCs or RARCs are attended by all interested administrations. The policy output of the conferences are regulations, amendments, modifications, deletions, footnotes, reservations, and resolutions constituting the ITU Radio Regulations.

These constitute only about nine percent of the total ITU "Arrangements" — the policy outputs of the other ITU organs, consisting of the Radio Recommendations, Telecommunications Services Recommendations, Telecommunications Services Regulations, and Convention and Administrative Council Resolutions. The broad range and sheer volume of technical standards and issues covered by the Radio Regulations are vividly illustrated by those approved at the 1979 WARC Final Acts. These consist of over 1100 pages of provisions, governing all aspects of radiocommunications – from operating practices, call-signs of stations, to, most importantly, the allocations of frequencies to specific radio services.

1. Allocations, Allotments, Assignments

The ITU spectrum policy-making process utilizes three methods for linking

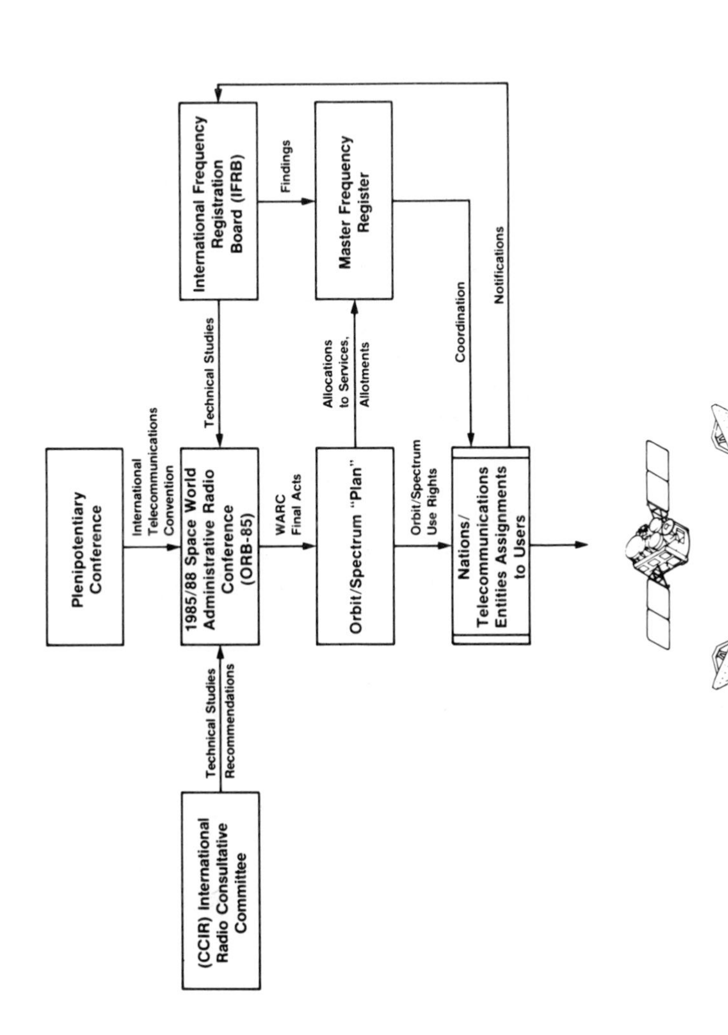

Table 4.1
The Orbit/Spectrum Policy Process

rights of access to use or users. On the international level, the ITU WARCs *allocate* frequency bands to radio*services*, both globally and by region. On the state level, countries or their telecommunications entities, designated "administrations" in ITU parlance, *assign* frequencies to stations or users within their boundaries or on ships, airplanes, and spacecraft registered under that country's laws. With the advent of satellite communications, an ITU WARC may also elect to *allot* frequencies and orbital slots not to services, but to individual states or groups of states for their use in providing the particular service specified by the WARC [5].

The terms allocation, assignmemt, and allotment allude to specific outcomes rather than simply categories or stages of policy-making. They are methods for administering access rights to use, which results in different distributions of access and use of the orbit/spectrum resource. They represent the central issues politicizing the access issue within the ITU. On the one side, the LDCs are demanding reforms that would mandate *allotments* of the orbit/spectrum resource to countries, while on the other side, the HDCs wish the ITU to remain an *allocator* of spectrum to services.

2. Jurisdiction of a WARC

In this respect, WARCs' jurisdictions vary. The 1959 and 1979 WARCs were so-called "General WARCs" which reviewed allocations to services over the entire radio spectrum — from 10 kHz to 300 GHz. On the other hand, the majority of WARCs have much more restricted jurisdictions, confined to specific services, regions, and frequency bands. For example, the 1971 WARC-ST (ST = Space Telecommunications) allocated frequency bands to space services. But while the 1977 WARC-BS (BS = Broadcasting Satellite) was convened for only the Broadcasting Satellite Service, it legislated *allotments* of frequencies and orbital slots to states.

During a WARC, decisions are reached by majority votes; each nation, regardless of its size or telecommunications use, may cast one vote. The Radio Regulations, once ratified, are recognized as codified international law. WARCs can be convened by Plenipotentiary Conferences or by other WARCs. Their scheduling and administration are performed by the General Secretariat and Administrative Council [6].

Inputs to the WARCs: CCIR and IFRB

In the ways that the CCIR and IFRB provide much of the technical information needed in order to achieve agreement at WARCs on allocations, standards, and procedures, they are also major points of interest for anyone seeking to influence the ITU policy-making process. The technical information is presented and discussed at conference preparatory meetings (CPMs) that are usually held during the year preceding the WARC. While both provide technical information, each organ has its distinctive constituency and role regarding frequency and orbit management. The CCIR and IFRB coexist as administratively semi-autonomous organs but with coinciding and occasionally conflicting duties and jurisdictions. While the WARCs are often big and flashy with growing media attention, the real work has been done long before the conference by small committees and study groups led by even smaller numbers of technical and industry experts "who know the right stuff."

1. The International Radio Consultative Committee (CCIR)

In theory, the CCIR performs a chiefly legislative function in the promulgations of Recommendations at CCIR Plenary Assemblies. Actually, the CCIR's main investigative and policy-making work is done in Study Groups which are assigned technical questions concerning certain areas of radiocommunications. There are 11 Study Groups and approximately 25 interim working parties.

In the area of orbit/spectrum access and use, Study Group 4, researching "Fixed service using communications satellites," and interim working party 4/1, studying "Technical considerations affecting the efficient use of the geostationary orbit," are the chief sources of technical information, Recommendations, and Reports to the 1985/88 Space WARC. Interim working party 4/1 has contributed the major CCIR research effort toward the 1985 Space WARC. Entitled, "Technical Report on the Geostationary Orbit," the 400-page volume contains detailed findings pertaining to how satellite systems function in an increasingly complex and crowded electromagnetic and orbital environments [7].

The word "Committee" in "CCIR" indicates its legislative function within the overall ITU-WARC organizational structure. The CCIR's functional relationship to the WARC resembles that of a committee involved in "marking-up" bills which are then sent on to the full legislative chamber, in this case, the WARC, for final passage.

The CCIR also has its own jurisdictional turf outside the WARCs. CCIR Study Groups and interim working parties conduct studies and hearings to formulate Recommendations and Reports on technical standards which are formally approved in Plenary Assemblies held every four years. The CCIR Recommendations, although legally non-binding, nonetheless enjoy near universal compliance for they represent the much-needed international standards for telecommunications equipment sold and used by a growing and diversifying world market [8].

The CCIR Study Groups are unique in that they hear and discuss a wide variety of viewpoints from not only members of official national delegations but also from private companies involved in telecommunications, designated Recognized Private Operating Agencies (RPOAs).

Once approved by the CCIR, RPOAs are allowed to work with and advise the Study Groups on technical standards for certain technologies. In return, the RPOAs contribute toward the operating expenses of the groups. United States RPOAs contribute two percent of the total ITU budget [9].

Many observers contend the arrangement is mutually beneficial. The CCIR Study Groups receive the latest technical information from the companies conducting telecommunications research and actually putting the technology into operation. The RPOAs gain in two ways. First, they have an input into the policy-making negotations concerning technical standards that either conform to their own product line or which they can fight because it would drive buyers to other suppliers. Secondly, the RPOAs by being privy to the information being swapped between negotiators also gain insights as to future technical standards and orbit/spectrum policies, which, in turn, affect technology markets.

In an important geopolitical sense, the RPOAs serve as evidence of HDC pre-eminence in the CCIR policy process. Out of 22 RPOAs from nine countries, only two from Kenya and Nigeria, represent LDCs. In addition to the RPOAs, 38 scientific and industrial groups along with 30 international organizations are approved participants in the CCIR Study Groups [10].

The relative absence of LDC representatives at CCIR Study Groups is seen as a serious disadvantage given the near-universal compliance CCIR Recommendations enjoy. Dr. Bruce Lusignan, Director of the Stanford University Communications Satellite Planning Center, describes the CCIR policy-making process as "dominated" by RPOAs and developed countries' Post, Telegraph and Telephone ministries (PTTs) who see enormous technological and economic interests at stake in the standards and recommendations [11].

As briefly mentioned in Chapter 1, Lusignan's comments are intriguing for many aspects of the communications satellite debate; especially as it revolves around the issue of orbit/spectrum resource scarcity. Lusignan argues that it is in RPOAs' and PTTs' economic interests to lobby for the adoption of stringent technical standards so that buyers are compelled to invest greater sums for the HDC-produced equipment that meets those performance standards. What the RPOAs seem to be saying, then, in CCIR Study Groups (and in the United States Space WARC Advisory Committee Meetings) is that the orbit/spectrum resource is indeed scarce, therefore justifying the technical standards they want. Thus, the United States and other developed countries, who continue to argue against perceptions of resource scarcity as a reason for detailed planning in their space WARC posturing, nonetheless tacitly accept such perceptions in their technical standards. In other words, such strict standards would not be necessary were the orbit/spectrum resource not as crowded and scarce as they purport in official policy statements [12].

In sum, although the CCIR decision-making bodies are extremely important in the drafting of the standards and recommendations which enjoy near-universal compliance, as well as affect billion-dollar markets, the process is also dominated by the space and telecommunications powers. As Codding points out:

> [A] most important characteristic of participation in the work of the consultative committees is the predominance of the developed countries, which is visible in the attendance at plenary assemblies but more striking in the working groups. At the 1978 CCIR Plenary Assembly, 19 out of 27 of the OECD-related countries were represented, as were six out of the ten Warsaw Pact countries. Only 36 countries from the Third World were represented out of a possible 117 . . . In the Study Groups, of the 39 countries which participated . . . only 14 were from the Third World: Argentina, Bahrain, Brazil, China, Korea (R.), Cuba, India, Indonesia, Iran, Nigeria, Senegal, Uruguay, Venezuela, and Yugoslavia. Of the 14, only six attended more than half of the 13 Study Groups . . . The countries with perfect attendance were primarily from the developed countries of the world: Australia, Canada, China, France, Germany (F.R.), India, Italy, Korea (R.), New Zealand, Nigeria, U.K., and U.S.A. [13]

However, the LDCs are finding out where the power lies in the ITU organization and are using their voting majorities to effect changes that could maximize their inputs. The administrative machineries of the CCIR and CCITT are each run by a Director, who was, before the 1982 Nairobi Plenipotenitary Conference, elected at each CCI's Plenary Assembly. At the Nairobi Conference, this changed. There, the LDCs succeeded in passing a proposal moving the site of the elections to the Plenipotentiary Conference. In this way, the LDCs can take advantage of their attendance at ITU Plenipotentiary Conferences by wielding voting majorities to elect the Directors, a power likely to be used as a bargaining chip for votes on other policy issues [14].

2. *The International Frequency Registration Board (IFRB)*

The five-member IFRB, created in 1947, performs duties as stipulated in Article 10 of the ITU Convention and Articles 9-15 of the Radio Regulations. Briefly, the duties are two-fold: First, to record frequency assignments made "by the different countries and of the positions assigned by them to their geostationary satellites," and secondly,

> to furnish advice to members of the Union with a view to the operation of the maximum number of radio channels in those portions of the spectrum where harmful interference may occur and to the equitable, effective, and economical use of the geostationary satellite orbit. [15]

The growing influence of the LDCs and their demands to change the telecommunications use regime are seen in the measure adopted at the Nairobi Plenipotentiary extending the IFRB's range of duties to include:

> [T]he needs of Members requiring assistance, the specific needs of developing countries, as well as the special geographical situation of particular countries and to provide technical assistance in making preparations for and organizing radio conferences . . . and assistance to the developing countries in their preparations for these conferences. [16]

The IFRB differs from the CCIR in the judicial focus of its definition of "harmful interference" and "international legal protection" of assignments:

> [I]t can be said that one of the IFRB's main tasks is to decide whether radio frequencies which countries assign to their radio stations are in accordance with the convention and the Radio Regulations and whether the proposed use of the frequencies concerned may cause harmful interference to other radio stations which are already in operation. Thus, the Board determines — on purely technical bases —the right of an Administration to use a given frequency for a specific purpose and the responsibilities the Administration thereby assumes vis-a-vis other Administrations. [17]

The IFRB is, through its registration and coordination procedures, the final arbiter of rights to frequencies and, concomitantly, orbital slots. Registration in the Master Frequency Register constitutes "legal recognition" of a particular state's use of a frequency, according such use "legal protection from harmful interference." [18] Although the IFRB is powerless to enforce its findings and must even enter those assignments made by administrations not in accordance with the Table of Frequency Allocations, IFRB registration nonetheless carries a great deal of weight due to the universal interference-retaliation sanction in the spectrum's "state of nature."

To a large extent, the IFRB acts as the global information clearinghouse, recognizing that the major problem in managing commons resources is that the over 158 users have to know what the others are doing, or planning to do. Not to trivialize the IFRB's work, but this is precisely its role since the spectrum's retaliation sanction takes care of the enforcement duties. One highly (if not over) qualified IFRB counselor showed me a pile of papers at least one-meter high lying on the floor of his Geneva office, saying, "These represent over 50,000 frequency assignments by country [X] because they are running a candidate for the Chair of Study Group [Y]. It's my job now to tell everyone else about them."

Briefly, an administration wishing "international legal protection" for its frequencies and orbital slots must undertake a three-step IFRB procedure consisting of advance publication, coordination, and notification of findings.

3. IFRB Publication, Coordination, and Notification Procedures

The IFRB publication, coordination, and notification procedures are stipulated by Articles 11, 12, and 13 of the Radio Regulations. To briefly sketch the outline of these procedures, an administration establishing a space satellite system must send the technical details of the system to the IFRB no earlier than five years and no later than three years before the system's initiation of operation. The IFRB circulates this information to the other ITU members.

Upon receipt of this information, other administrations with either already operating or planned satellite systems and who suspect that the new system could cause "harmful interference" to their own, are directed to contact that country and inform it of their concerns. Upon hearing the complaints, that country now must either find ways to re-design its system so that it will be compatible with those already operating or registered, or argue that the criteria set up by the ITU and IFRB are unrealistic. Affected countries enter into discussions on ways to minimize harmful interference with assistance from the IFRB.

Following these talks, the IFRB performs a coordination procedure consisting of a technical examination of the systems in question, so as to ascertain their interference compatibility.

If the system has been found not to produce "harmful interference" to other users above IFRB standards, the newcomer administration may so notify the IFRB which duly records the parameters of the new system in the Master Frequency Register. The IFRB Circular thereupon advises other administrations as to the new system's legal status [19].

The Master Frequency Register, then, is the listing of all frequency assignments and orbital locations published and, if necessary, coordinated with the IFRB by complying administrations. It is the legal "final step" in achieving access to the orbit/spectrum resource for satellite communications. It is also the source of much of the dissatisfaction over the method for allocating rights to orbital bandwidth.

> The satisfactory completion of the coordination process entitles an administration to notify its planned space system to the IFRB which will ... enter the date of receipt of the notice in the Master International Frequency Register ... Generally, the registering of system frequencies in the Master Register indicates that "international recognition of the use of the frequencies" had been obtained on a specific date; *the onus for coordination is then placed on systems notified at a later date in the Master Register. This is the basis of the "first-come, first-served" principle.* [20]

4. Assessment of IFRB Coordination and Notification Procedures

Economically, as well as technically, the latecomer finds coordination and notification procedures biased considerably in favor of already operating or notified satellite systems. These procedures are also out of phase with the initiation, financing, design, development, and final prepartions for the launching of a communications satellite, a process that typically requires up to ten years.

In that time span, many fundamental technical and economic decisions are reached based on assumptions of allowable performance levels given a certain population of satellites in the intended portion of orbital arc. This is the chief criticism of Fabrio Galante, Systems Planner and Analyst for Eutelsat, who enunciates the latecomer's plight as follows:

> It is noteworthy that since about 1980 *all* newly published systems operating at 4/6 GHz had to go through the coordination process because the 4% criterion [of added interference to other systems] was exceeded in all cases; it is also worth noting that for the 14/11 GHz frequency bands there are only about 80 out of 360 degrees of the orbit where it would appear that new systems, using these bands, can be operated without going through the coordination process.
> Advance publication of a new system has to occur no earlier than five years and no later then two years before the system becomes operational. It usually takes a minimum of one year from the mailing-date of the Advance Publication Information and the beginning of the coordination process.
> If we compare these periods with the usual time needed to establish a system . . . we can immediatly see that an Administration will initiate the coordination process when all its system parameters are practically frozen and all equipment is procured. Consequently, *any* change in the characteristics of the system resulting from the coordination *will have an economic impact* [emphasis added]. [21]

In sum, the timetable for IFRB notification and coordination does not match the design and construction schedule for modifying new satellite systems at the lowest cost to the newcomer. The changes to satellite design and performance often result, in Galante's view,

> in a considerable reduction of the system's economic viability because the system operator has to compromise on parameters that the system planner has optimized generally with no regard to the environment [interference in geostationary orbit and to coverage areas] where the planned system will have to operate. [22]

Galante's perspective of the IFRB's coordination procedures, being that of the satellite system planner, is invaluable for understanding the discussion on the following pages concerning orbit/spectrum planning and *a priori* rights. In his critique, Galante has come the closest to defining orbit/spectrum scarcity. Scarcity exists whenever calculations show that the new system would exceed the four percent limit on additional interference to other systems, both operating and notified. At that point, the newcomer must undertake the coordination procedure, while the already operating or notified system is under no legal obligation to change its parameters to accomodate the new system. Satellite system planners are aware of these technical and economic implications and it is their input to the policy-making process which has fueled the debate over the method for allocating the orbit/spectrum resource.

5. IFRB Inputs to the WARCs

The IFRB provides three types of inputs into the WARCs: (1) technical information, (2) the Master Frequency Register (which lists frequencies and orbital locations of geostationary satellites), and (3) a policy history of findings and coordination procedures upon which affected states base many of their policy positions.

As the disseminator of technical information, the IFRB sponsors Panel of Experts (POE) meetings during the year preceding a WARC. In preparation for the 1983 Regional Administrative Radio Conference on satellite broadcasting for North and South America (RARC-83 for ITU Region 2), the IFRB sponsored a series of Panel of Experts meetings. The last POE meetings, held in December 1982 at the Geneva Conference Center, reviewed and refined computer programs designed to greatly simplify testing and predicting the interaction of broadcasting satellite systems. This was no small task given the extreme complexity and variance in configuraions of antenna patterns, satellite footprints, transmitter power levels, and receiver characteristics. The computer models allowed various proposed orbit/spectrum plans to be tested for electromagnetic compatibility, and, at the same time, served as indices for efficient use and equitable access criteria — crucial bargaining chips during conference negotiations.

United States delegate to the RARC-83 POE, Edward Jacobs, emphasized over coffee on a wet, foggy Genevan morning, the usefulness of the computer models to the successful outcome of the conference:

> The POE allows participants to gain first-hand experience with the computer program so that delegations can go into the RARC with confidence in the validity of the computer program and its findings. Perhaps more importantly, the POE meetings, as well as the CCIR Conference Preparatory Meetings, serve the invaluable function of bringing all delegations to a workable minimum-common level of technical expertise regarding the complex engineering considerations involved in devising orbit/spectrum management and planning arrangements; such a "common language" is necessary in order to accomplish the conference's objectives in the designated time span. [23]

The second major IFRB input to the WARCs, the Master Frequency Register, records those frequency assignments and orbital locations granted international recognition and legal protection from harmful interference, and all others by date of entry even if they did not qualify for full recognition. The Register's listings also serve a political function by indicating relative numbers of spectrum and orbit assignments between HDCs and LDCs (as discussed in Chapter 2). However, the validity of the Register is limited by the Board's inability to conduct independent audits or to monitor actual spectrum and orbit usage. It must rely instead on the published notifications submitted by administrations voluntarily complying with Articles 11 through 13 of the Radio Regulations [24]. States have been reluctant to grant the IFRB the authorization to delete assignments from the Register given the priority rights afforded the first "user." In recent years, the IFRB was directed to conduct a review of the hundreds of thousands of listed assignments so as to clear out the "deadwood entries," mostly found in the longest-used High Frequency shortwave bands. Needless to say, these debris-clearing procedures themselves take years to complete.

Extension of the ITU Regime Out to the Geostationary Orbit

The ITU regime encompasses space telecommunications by way of its role as regulator of the spectrum — its traditional and successful area of jurisdictional competence. Spectrum exists in the space environment just as it does on earth, and that the ITU should also regulate and manage its use by artificial earth satellites was never seriously questioned. The same cannot be said for the geostationary orbit.

As we have seen, the orbit is an extremely desirable area of outer space. As discussed in Chapter 2, geostationary "parking slots" are actually a group of flightpaths through a certain region of space in which intersecting laws of Newtonian orbital mechanics permit satellites to appear "fixed" in an earth-bound observer's sky. While the Bogota Declaration seeks to extend sovereign airspace to geostationary altitudes, it has not successfully swayed international legal opinion nor the customary practices of states who recognize outer space to be a *res communes*.

The ITU has become the international organization most directly involved in policy questions dealing with the geostationary orbit. The ITU has assumed this added area of jurisdiction through the organization's historical function of managing *spectrum allocations* for geostationary satellites. While one can argue that the ITU's long-standing success as an international organization stems from the tight fit between the characteristics of the spectrum resource and its decision-making process (federal structure, committee system, RPOAs, *et cetera*); can the organization also accommodate the geostationary orbit onto its spectrum-based institutional structure? In other words, how much of the current politicization of satellite communications can we trace to unintended consequences stemming from the ITU's assumption of geostationary jurisdiction?

How Did the ITU Get into Space?

1. WARC 1959 and EARC 1963

At the 1959 General WARC, held in Geneva, the ITU for the first time allocated frequency bands specifically for space radiocommunications services. The successful flights of the Sputnik, Explorer, and Vanguard vehicles made it clear that space was to be actively explored with spacecraft using frequencies for a wide range of functions from guidance and telemetry to voice and television links, each service presenting its own spectrum requirements. The 1959 WARC did not find itself competent to deal with these topics and called for an Extraordinary Administrative Radio Conference (EARC) for the space services, to be held in 1963 [25].

The EARC met in Geneva during October 1963. The conference revised the Table of Frequency Allocations, allocating a total of 6076.462 MHz for the various space services on either a shared or an exclusive basis. Of these allocations, 2800 MHz were designated for communications satellites on a shared basis with other terrestiral services. This marked a sizable increase in allocated spectrum for space services, rising to 15 percent from just 1 percent of all frequencies allocated at the 1959 WARC. In addition, the EARC recommended convening another Extraordinary Administrative Radio Conference to review the rapidly developing techniques and relevant regulations and allocations for space services.

The EARC also passed a Recommendation stating the need for CCIR studies regarding "the use of satellite transmissions for direct reception by the general public of sound and television broadcasts." Thus, DBS was brought into the ITU fora at a date when the first prototype geostationary satellites were just being launched. The EARC delegates witnessed the capabilities of satellite communications first-hand when the Geneva meeting was linked via Syncom II to Washington and New York [26].

Most importantly, in a policy sense, the conference adopted a Recommendation recognizing

> that all Members and Associate Members of the Union have an interest in and right to an equitable and rational use of frequency bands allocated for space communications "and" that the utilization and exploitation of the frequency spectrum for space communication be subject to international agreements based on principles of justice and equity permitting the use and sharing of allocated frequency bands in the mutual interest of all countries. [27]

2. WARC-ST

WARC-ST (ST = Space Telecommunications) was held in Geneva during June and July 1971 [28]. The conference marks the officially-recognized extension of the ITU regime to the geostationary orbit. In effect, the ITU was "catching up" to technology's rapidly expanding utilization of the orbit/spectrum resource in ways and with problems largely unforeseen at the EARC eight years earlier. Several milestones were reached in 1969. One week before Neil Armstrong and Buzz Aldrin became the first human beings to visit another celestial body, a third INTELSAT satellite was backed into its parking slot over the Indian Ocean. This completed the first truly-global telecommunications network just in time to provide live television coverage of the moon walk "via satellite" to most inhabited areas of the moon's planetary neighbor.

Two years later, the 1971 WARC revised the Radio Regulations, formally extending ITU jurisdiction to the geostationary orbit. WARC-ST produced several significant agreements regarding use rights to the orbit/spectrum resource, the most important of which is Resolution Spa 2-1:

> Relating to the Use by all Countries, with equal Rights, of Frequency Bands for Space Radiocommunication Services
> [A]ll countries have equal rights in the use of both the radio frequencies allocated to various space radiocommunication services and the geostationary satellite orbit for these services . . . [Furthermore], registration with the ITU of frequency assignments for space radiocommunication services and their use *should not provide any permanent priority* for any individual country or groups of countries and should not create an obstacle to the establishment of space systems by other countries [emphasis added]. [29]

As Christol writes,

> This language was a clear rejection of the view that a registration with the IFRB of a national assignment of a radio frequency would establish a permanent priority for the registrant respecting the use of a particular frequency. [30]

Resolution Spa 2-1 recognizes and reinforces the *res communis* status of the orbit/spectrum resource. A state's notification to the IFRB and subsequent registration in the Master Frequency register provides international recognition and legal protection for the *engineering lifetime* of the satellite. In other words, the right to international protection for a particular orbital slot and frequencies cannot be extended past the nominal operating lifetime of the satellite originally notified. A state may not "appropriate" an orbital slot in a *de facto* manner by replacing spent satellites with new ones [31].

The conference also recognized different technical requirements and interference problems associated with diverging types of satellite services, particularly with broadcasting satellites and their effect on low-power telecommunications satellites. The higher power levels of broadcast satellites, necessary to transmit programs to small rooftop antennas, raised far-reaching questions regarding their electromagnetic compatibility with the low-power, connectivity satellite services. In order to mitigate potential interference, the conference established two space services, the Broadcasting Satellite Service (BSS) and the Fixed Satellite Service (FSS), and allocated specific frequency bands to the BSS. The significance of this step is underlined by Rothblatt:

> [T]he 1971 Conference prepared the way for greater future international control by defining a new radio service, the Broadcasting Satellite Service, allocating frequencies for this service on a worldwide basis, and resolving that this new service be established as part of a comprehensive plan. Because broadcasting satellite service, like any space service, ultimately depends upon a frequency linked to an orbital position, the ITU would inevitably plan the use of the vast geostationary orbit — a truly quantum step for international regulation and a correspondingly great sacrifice of national freedom of action. [32]

The conference placed the space radiocommunication services under the jurisdiction of the IFRB. Nations were required to notify the Board concerning:

1. Basic operational data on any planned satellite system,
2. Coordination of foreign administrations in eliminating potential interference, and
3. Designing systems in a manner that would minimize satellite transmissions over foreign territory. [33]

While the conference "rejected giving the power to the ITU to allocate orbital positions[, it] did, however, assert the power to record or register orbital positions ... assigned by member States ... " [34] WARC-ST, by asserting the ITU's jurisictional competency to regulate and manage the geostationary orbit, marks the beginning of the ITU's involvement in the larger controversy over rights to *space* resources, a debate encompassing many more issues and negotiating fora. The legal, political, and technical origins of the conflict are well summarized by Christol, who writes:

> These resolutions . . . although without legal force, have had a critical impact on and have given direction to subsequent ITU decisions. The ongoing contest between states possessing a present telecommunication capability in space, and those not possessing such a capability, has been identified with both present and potential future uses of frequencies and orbital positions. The space-resource states have

> wished to be able to make a present use of their scientific and technical capabilities. The other states have tried to protect their claims respecting the future free and equal exploration and use of and free access to the space environment so that they might exploit their own space potential. States having the present potential have opposed allocations to states which would not be able to make use of them. States not having a present potential have objected to the use by the space-resource states of space environment resources in such a way as to preclude their later sharing in such uses and benefits.
> Nonetheless, the consensus emerging from the 1971 meeting was that states should be accorded present opportunities to explore, use, and exploit such resources, including preferences encompassing priorities respecting uses of frequencies and orbital positions, even though these uses were not to be considered exclusive rights appertaining to the user. They were conceived of as rights to use but not ultimate proprietary rights. Hence, there was a specific expectation that over time distributions of opportunity would be realized so that an equitable sharing of resources might result. At the same time, the traditional claim of "first-come, first-served" was generally disavowed. [35]

Here we see the ITU attempting to define access rights to a new global common property resource — rights caught between contradictory premises of "free access" and "equal rights to use." To a great extent, the conflict lies in the nature of the resources themselves. In terms of access, the geostationary orbit is a resource quite unlike the radio spectrum. While the spectrum is universally accessible, the geostationary orbit is not. In 1971, only the United States and the Soviet Union possessed the technical capabilities and financial resources to launch satellies into a geostationary flight path. Thus, while the universal accessibility of the spectrum conformed to the previous bases of ITU rights and rules, the geostationary orbit did not. The controversy stems from a basic incongruency between configurations of capabilities and interests of states possessing the technology, and the decision-making process dominated by a majority of non-space powers.

Resolution Spa 2-1, attempts to deal with the problem by asserting that notification and registration by the IFRB conferred no permanent priority to use, a rule in alignment with limited satellite engineering lifetimes. This represents a major change in the concept of "priority rights" to the first user; space radiocommunications were not to be governed by the same priority of use rules that had been historically followed for terrestrial spectrum users.

1973 ITU Plenipotentiary Conference

The ITU held its 1973 Plenipotentiary Conference in Torremolinos-Malaga, Spain. The legislative product of the conference, the International Telecommunication Convention, formally defined the status of the geostationary orbit and spectrum resources for space radiocommunication in Article 33:

Article 33

Rational Use of the Radio Frequency Spectrum

and of the Geostationary Satellite Orbit

1. Members shall endeavour to limit the number of frequencies and the spectrum space used to the minimum essestial to provide in a satisfactory manner the necessary services. To that end they shall endeavour to apply the latest technical advances as soon as possible.
2. In using frequency bands for space radio services Members shall bear in mind that radio frequencies and the geostationary satellite orbit are *limited natural resources*, that they must be used *efficiently and economically* so that countries or groups of countries may have *equitable access to botch* in conformity with the provisions of the Radio Regulations according to their needs and the technical facilities at their disposal. [emphasis added] [36]

In order to implement the provisions of Article 33, the Convention modified and added to the essential duties of the IFRB as specified in Article 10:

b. to effect . . . an orderly recording of the positions assigned by countries to geostationary satellites;

c. to furnish advice to Members with a view to the . . . equitable, effective and economical use of the geostationary satellite orbit;

d. to perform any additional duties, concerned with the assignment and utilization of frequencies and with the utilization of the geostationary orbit, in accordance with the procedure provided for in the Radio Regulation, and as presecribed by a competent conference of the Union . . . [37]

By 1973, the lines of debate over the geostationary orbit and spectrum resources were being drawn as states realized the economic, political, and military value of satellite communications. A battle over access and use rights to the orbit/spectrum resource was being fought in the ITU, ostensibly over how to devise a management scheme that would ensure equitable access with efficient resource utilization. Curiously, the mid-1970s was also a period of time in which the geostationary orbit was relatively empty except for some 20 to 30 non-interfering satellites. Were states, then, really concerned about orbital saturation, or were there other policy objectives driving the accelerating debate on towards its 1979 confrontation? One example of how use concerns dominate access objectives took place in 1977.

WARC-BS (1977)

The World Administrative Radio Conference for Broadcasting Satellites (WARC-BS) was held during January 1977 in Geneva. The Conference, pursuant to the authorizing resolution approved by WARC-ST, had the task of planning the Broadcasting Satellite Service in the 12 GHz (Ku) Band [38]. A World Agreement and Associated Plan were "amicably adopted" for Regions 1 and 3.

The plan specifies the number of channels and orbital slots as well as the accompanying technical parameters for the operation of direct broadcast satellites in great detail. The number of channels vary from two for Brunei and other small states to sixty-four for the Soviet Union, while most countries received four, "depending on size, population and foreseeable communication needs." [39] The political significance of the WARC-BS plan was immediately apparent due to the unmistakable indication that about two-thirds of the world's countries preferred to "sacrifice efficiency in order to gain more equitable access." [40] The plan was a victory for the proponents of *a priori* resource allotment and management. All countries, whether or not they even had a long range telecommunications goal incorporating satellites broadcasting, received orbital slot and channel allotments. In other words, the countries gained the *rights* to these portions of the orbit/spectrum resource, however, they did not gain the *means* for achieving technical access to them.

This again raises the question of a state's policy motivations and how they affect the configurations of issues, interests, and capabilities of the ITU regarding use and access management of the orbit/spectrum resource. Why did so many diverse states, from the least-developed African and Asian states to the highly-developed space-resource states of Western Europe and the Soviet Union, agree to this arrangement?

The answer lies with the conflict between the technological capabilities of the Broadcasting Satellite Service and national sovereignty. Technical constraints intrinsic to the design of spacecraft antennas prevent the fitting of a signal footprint exactly within the borders of a target country. Instead, there are varying amounts of signal "spillover," depending on the size and shape of the country or service area, the proximity of neighboring foreign population centers, and most importantly, the technical performance of receiving earth stations [41].

For example, a Swiss direct-broadcast satellite aimed at Switzerland would also illuminate most of France, Italy, the Federal Republic of Germany, and Austria with whatever programming the Alpine state wished to transmit. This technical breakthrough in continental broadcasting from a single satellite, conjuring up visions of American-style "sitcoms" being sent from a possibly privately-owned and operated satellite to vulnerable school children in Social-Democratic Western Europe (not to speak of Communist Eastern Europe), aroused many states' maternal fear for their national telecommunications sovereignty.

It is interesting to note the effect that the "radio wars" in Europe — being fought since the mid-1960s between government-operated radio networks and pirate broadcasters in the North Sea — may have had on the ITU negotiations. The clandestine operations, transmitting what state-owned and controlled radio in Europe banned — non-stop rock music — were a highly visible nuisance, exposing unsuspecting teenagers to programming frowned upon by many governments who also resented the diversion of advertising revenues to North Sea trawlers. The pirates transmitted from ships outside territoral waters in the North Sea; or as in the case of the station that called itself the "Big Noise in Europe" — Radio Luxembourg — which was state-licensed broadcasted from its vantage point in the middle of the continent, effectively blanketing literally all of Europe with commercial-style programming outside its neighboring states' control.

Of course, a satellite in outer space is even farther away from any means of state control. In light of this prospect, the issue of "prior consent" provided world policy-makers with perhaps the strongest incentive for the adoption of a detailed orbit/spectrum plan, and not, as stated by these countries, their concern for interference. This finding is borne out by the fact that as of mid-1985 — eight years after the conference — there are no direct broadcast satellites using the WARC-77 plan [42].

The term, "prior consent," refers to an arrangement whereby, before a state may transmit programming from its satellite that can be received in other countries, that state must acquire the consent for such broadcasting from the affected neighboring countries. A resolution outlining the rights and responsibilities of states regarding such prior consent arrangements has been on the COPUOS negotiating table since the mid-1960s, when it was first feared that satellite broadcasting and concomitant loss of national sovereignty were imminent.

This perception was moderated when it was realized that there are technical means to prevent one's own population from receiving unauthorized satellite broadcasts. In fact, the problem today is how to make the various satellite systems electronically compatible so viewing audiences in Western Europe will be able to receive programming using different transmission standards (as in PAL and SECAM, the West German and French color television standards, respectively), thereby enlarging both the audience and revenues to pay for the expensive satellites. The European Broadcasting Union and the individual manufacturers are discussing the merits of various modulation techniques; at stake are large investments in technology for a particular method and equipment support for already initiated systems [43].

Nonetheless, many states have insisted upon their sovereign "prior consent" prerogatives. Under almost all the proposals for a "prior consent" regime, though, the sovereign rights of one state to transmit programming come into conflict with those of another state to control what types of programming may come in over its border. Should an outside state have the prerogative to determine what a state may broadcast to its own people? The continuing impasse over the issue prompted many states to agree to the technical parameters placed into the 1977 Conference Plan. As Codding states:

> Many socialist and Third World countries saw the adoption of a plan for the broadcasting-satellite service as an opportunity not only to obtain their "fair share" of the radio resource, but also as a means to circumvent the prior consent stalemate. The adoption of a plan for providing domestic DBS service would effectively preclude one country intentionally broadcasting into another. [44]

Technical Consent

As we have seen, widespread concern over satellite broadcasting's threat to national information prerogatives prompted the passage of Paragraph 2674 at the 1971 WARC-ST (in anticipation of the 1977 WARC-BS):

> In devising the characateristics of a space station in the broadcasting satellite service, all technical means available shall be used to reduce, to the maximum extent practicable, the radiation over the territory of other countries *unless an agreement has been previously reached with such countries*[emphasis added].

In addition, Paragraph 2612 states:

> Space stations shall be fitted with devices to ensure immediate cessation of their radio emissions by telecommand, whenever such cessation is required under the provisions of these Regulations. [45]

The technical coordination required by Paragraph 2674, in the view of many observers, constitutes a type of "technical consent," which although consistent with previous ITU rules and allotment plans for other frequency bands, is for satellite broadcasting a form of "prior consent." ITU Secretary-General Butler commented in an interview a few weeks after his election to that post, that Paragraph 2674 accomplished with a technical procedure what many states have been unable to pass as legislation in other fora, notably COPUOS [46].

In sum, the 1977 Plan represents a compatible match-up between the odd couple of technology and national interests, or to put it another way, access and use. In terms of technology, the heightened interference potential posed by broadcasting satellites requires greater coordination between users. In order to make the system economically feasible, the home receivers must be inexpensive, necessitating small antennas with less discriminatory power. The small antennas are satisfactory only when used with very powerful satellite transmitters, separated far enough from each other in the geostationary orbit so that receiving antennas do not receive two DBS signals simultaneously on the same channels.

The combination of the two factors represents a severe interference potential, so that even with relatively few satellites, the orbit/spectrum resource appears saturated, especially when service areas are as geographically concentrated as in Europe. Therefore, in order to avoid interference problems, close coordination in the design and operation of the various satellite broadcasting systems is necessary between users.

Many countries place high priorities on maintaining their sovereign prerogatives to control the use of the spectrum within their boundaries, and most importantly, to control or regulate the content of the broadcasting that does take place. In this case, technology came to the aid of politics. In a strictly technical application of Paragraph 2674, political "prior consent" stipulations are, in effect, translated to the technological problem of reducing interference. The 1977 Plan was a "victory" for national prerogatives over the "free flow" of information, instituted as technical coordination requirements and, most importantly, stated in *technical language* [47].

In the Americas (Region 2), the grounds for a happy marriage between technology and politics did not exist. Throughout this period of time the United States opposed detailed planning for the 12 GHz band, both on the technical merits of the Plan and for the political implications such a plan would project as a precedent for later ITU conferences. However, "Planning" was promulgated for Region 2 at the 1977 Conference in the form of an arc segmentation plan. In order to reduce interference problems, WARC-BS divided the orbital arc over Region 2 into sections designated for low-power FSS and high-power BSS. This form of segmented allocation of the orbital arc was in place only two years until the 1979 WARC rescinded the segmentation provisions.

Significance of the 1977 Plan

The 1977 Plan represents the largest extension of jurisdiction by an international organization into any global commons. About two-thirds of the orbital bandwidth in the 12 GHz band was allotted to countries, thereby restricting open access. The plan unmistakably underlines the political reality that most countries felt disadvantaged by the old regime based on the principle of first-come, first-served. As Rothblatt explains:

> Under the traditional rules, the first nation to place satellites in orbit acquired "significant advantages" over others. Despite the express

> denial of any permanent priority and despite an affirmative obligation to make room for new systems, the newcomer remained in an inferior position, obligated to "approach the existing stake holders and seek such accommodation as they were willing to provide." Though ITU representatives patiently explain that the assignment of nominal orbital positions does not represent appropriation of space resources, which would contravene both the 1967 Outer Space Treaty and the ITU's jurisdictional authority, [48] the assignment plan does at least reverse the criteria by which preferential rights are determined in interference disputes involving 12 GHz broadcast-satellite services. Instead of requiring the newcomer to "seek such accommodation as [the first-comers] are willing to provide," *the superior bargaining position now rests with the state whose service conforms to the plan — whenever it plans to orbit its system and regardless of how long a nonconforming service has been in operation* [emphasis added]. [49]

To reiterate, the Plan is an agreeable meshing of technical and political interests. The technical constraints posed by powerful satellites and inexpensive home receivers require extensive user coordination, whether part of an *a priori* plan or not. The technical details of the Plan allowed a form of prior consent under the ruse of technical parameters to prevent interference. Finally, having allocated the frequencies in 1971 with the expectations of a great demand for satellite channels, by 1977 states were aware that the Plan was more than adequate in supplying each state's needs.

This change in attitude can be seen in West Germany, where the current debate over cable television is raising the question of whether viewers need more than the three current stations each region provides. Even in the United States, the once rosy economic picture projecting up to 100 cable channels has dimmed due to saturated demand and limited revenue base for already operating channels. Indeed, the major significance of the 1977 Plan was its symbolic value showing that the ITU's business-as-usual attitude regarding orbit/spectrum allocations was deeply mistrusted by a majority of the world. Although the United States was able to "limit the damage" to ITU Regions 1 and 3, the Plan was a setback to its policies favoring flexible planning and open access, giving that country a taste of the struggle to come in 1979.

World Administrative Radio Conference 1979

The international political environment and information technologies had both undergone transformations since the last General WARC in 1959 [50]. The specter of polarization between the North and South and the East and West, loomed on seemingly intractable issues as delegates gathered in Geneva in the fall of 1979. Brian Segal, a Canadian delegate, wrote:

> The context within which international decision-making occurs is . . . characterized by world tensions at one level and pressures for self-preservation and collective agreement at another. For three months in late 1979, in this atmosphere of conflicting pressures, over 2,000 delegates assembled at a conference in Geneva to establish a new set of world regulations governing the shared use of the radio spectrum and the geostationary satellite orbit. Mankind, unwittingly, is grateful that this conference achieved most of its objectives. If it had failed, the resultant anarchic use of the radio spectrum for broadcasting and communications could have brought unimaginable chaos to world society. [51]

The 2000 delegates and advisors attending WARC-79 represented 142 member countries. Significantly, almost one-half of the countries eligible to attend WARC-79 were not even independent states when the last General WARC was held in 1959. As Glen Robinson, United States Ambassador to WARC-79 commented, "[It] was the largest conference in the history of the ITU, in terms of the numbers of participants, proposals, and decisions made." [52]

Many Third World delegations came to Geneva with experienced negotiators, their skills honed during long bargaining sessions in other international fora within North-South contexts of the New International Economic Order (originating in the United Nations Conference on Trade and Development (UNCTAD) and the New World Information and Communication Order (first articulated in the United Nations Educational Social and Cultural Organization (UNESCO) [53]. Consequently, Third World delegations adroitly placed the supposedly "technical" issue of access to the orbit/spectrum resource squarely into the ongoing political debate over asymmetrical North-South resource distribution. For example, Third World spectrum experts were able to extract from the Master Frequency Register that the United States and Soviet Union representing only 15 percent of the world's population, use 50 percent of the spectrum; while another estimated that 90 percent of the spectrum is controlled by 10 percent of the world's population [54].

The dominant Third World perspective is highlighted by the following quote:

> Third World nations saw WARC as a struggle to obtain equitable access to spectrum and orbital slots from those who occupy them now. A July 1979 African Press Service release, speaking of African needs in the HF band or broadcasting and fixed communications, maintained that the "biggest users of this frequency are Britain,

> France, the Soviet Union and the United States. They use this frequency mainly for spying and propaganda purposes." [55]

The United States approach to the conference is illustrated by the following statement by Glen Robinson, written after the conclusion of WARC-79:

> It was widely predicted that WARC would be the occasion for a North-South confrontation over the principles of a new information order. Some thought such a debate desirable as well as inevitable and strongly criticized the US position that such a debate should be avoided. According to these critics, WARC was an opportunity, to be seized and not shunned, for the United States to take a stronger initiative to support Third World aspirations and needs. Such initiatives never were defined precisely, but the thrust of the critics' demand was that the United States should offer greater technical assistance and should support Third World proposals to obtain "more equitable access" to the spectrum . . . The US position was not hostile to considering such issues at WARC, insofar as they were relevant to the agenda . . . The United States had no illusions that the WARC would be purely "technical." There was never any doubt that "political" factors — in a broad sense of that term — would be critical, despite the ostensibly technical character of the subject matter . . . Purely technical cases are as uninteresting as they are exceptional. [56]

Against this backdrop of technical and political controversy, delegates labored under an unprecedented workload. Each delegate received copies of over 1,500 documents totaling more than 5,000 pages, available in the three working languages — English, Spanish, and French. The discussions and negotiations were conducted by nine committees which used over 120 working groups to consider over 15,000 submitted proposals, 12,000 of which concerned frequency allocations. It can be seen from the above statistics that a sizable delegation was needed by each country just to keep tabs on the work being done by the various committees and Study Groups.

The United States, for example, fielded a delegation of 67, accompanied by two Congressional advisors and a 40-member support staff. In addition, the United States delegation came prepared with a computer-satellite link to a data base in Washington which stored carefully pre-considered "fall-back" policy options in situations where opposing proposals won key votes.

The size and resources of most of the other delegations were much more modest. Of the 142 attending states, 87 delegations had less than 10 members, 39 had between only 10 and 20 members, 8 had between 21 and 30, while only 8 delegations numbered more than 30 [57].

Size, unavoidably, had an effect on any delegations's ability to shepherd its proposals through the conference's decision-making labyrinth, as Segal points out:

> [T]he complexity of the conference structure and the large differentials in delegation size reflect to some extent the disproportionate participation, control and influence of the developed countries in conference proceedings . . . [However, t]he size of a delegation is only one factor contributing to the amount and quality of input. The depth and breadth of technical knowledge of the Radio Regulations, of the different service technologies and requirements and of the complexity of international frequency matters, (e.g., propagation, a system's technical characteristics, frequency management techniques, space issues) had a major bearing on the ability of a delegation to contribute to all aspects of the work of the conference . . . Obviously, many developing countries were unable to participate actively in debates and negotiations at all working levels . . . many delegations were faced with voting on matters only in their final stages instead of participating in the evolution of solutions and compromises. [58]

In sum, the decision-making environment of the eleven-week conference reflected overall transformations in the international system brought on by the large influx of developing countries and tighter interdependencies between all states. The big users of spectrum and orbital slots sent large, well-prepared and financed delegations, while some developing countries, attending a General WARC for the first time, assigned small delegations without extensive preparation or expertise in telecommunications technologies or policies. In other words, the asymmetrical global distribution of national resources, both natural and manmade, was mirrored in the meeting rooms of the Geneva Conference Center.

WARC-79 Policy Outputs for Orbit/Spectrum Access

In the view of many observers and participants, the major issue of the conference concerned how to establish "rights vesting mechanisms," assuring equitable access for all states to the orbit/spectrum resource. Or more succinctly, how is the orbit/spectrum resource to be planned? On this crucial issue the conference was collectively unable, or unwilling, to decide and risk possibly destructive polarizations that could negatively affect agreements in other areas. The confrontation was defused by resorting to a postponement of the debate through Resolution BP. This is the most significant outcome of WARC-79 regarding the issue of access to the orbit/spectrum resource [59].

1. Resolution BP

Resolution BP, although legally non-binding, makes several important statements as reproduced in part below:

Resolution BP

Relating to the Use of the Geostationary-Satellite Orbit and the Planning of Space Services Utilizing It

a) [T]he geostationary-satellite orbit and the radio frequency spectrum are limited natural resources . . .

b) [T]here is a need for equitable access to, and efficient and economical use of, these resources by all countries as provided for in Article 33 of the International Telecommunication Convention, 1973, . . .

c) [T]he utilization of radio frequencies and the geostationary-satellite orbit by individual countries and groups of countries can take place at various points in time, based on their requirements and the availability of the resources at their disposal;

d) [T]here are growing requirements all over the world for orbital position and frequency assignment for the space services;

[The Conference] resolves:

1. that a world space administrative radio conference shall be convened not later than 1984 [subsequently changed by the Administrative Council to 1985] *to guarantee in practice for all countries equitable access to the geostationary-satellite orbit and the frequency bands allocated to space services*; [emphasis added]

Resolution BP goes on to stipulate that the conference shall be held in two sessions; the first to "decide which space services should be planned" and the technical parameters which would form the basis for the second session's work on implementing the arrangements into some kind of institutional framework [60].

Resolution BP also illustrates what WARC-79 did *not* do, namely, allot to states orbital slots and channels on an *a priori* basis. Such an action was clearly within the purview of the conference's charter and it was certainly paramount in the pre-conference policy positions of many developing countries.

2. The Controversy over Planning and Resolution BP

Resolution BP was the most hotly debated issue at WARC-79. It represents both the politicization of the ITU, and at the same time, the organization's strong tradition of cooperation and compromise for reconciling radiocommunications disputes. The significance of Resolution BP is as ambiguous as the issues surrounding the orbit/spectrum resource are complex. Indeed, ambiguity may be deemed necessary in the face of complexity, as the latter is often the cause of the former. But let us now look at Resolution BP from the perspective of two representative countries, the United States and India.

2.1. View from the United States

The United States is the most outspoken opponent of any WARC-BS form of detailed planning and is one of the only countries to consistently oppose administrative or legal restraints on information flows which detailed planning could introduce. While the concerns for prior consent prompted many Western European and Eastern Bloc countries to vote with the LDCs in approving the WARC-77 Plan for Regions 1 and 3, these same countries, in general, oppose detailed planning. The United States position, as argued by FCC Delegate Rutkowski before the Working Committee Six Ad-Hoc Two, summarizes the positions held by other HDCs on detailed planning:

> In deciding on the best planning approach, we trust that the future Space Conference will recall the grave problems foreseen by the United States, and also by a number of the countries who favored planning for the broadcasting-satellite service in 1977, if 1977-type planning were to be applied to the fixed-satellite service . . .
>
> *Accommodation of Long Range Requirements Predicted With Near-Term Technology*
>
> A detailed orbit-frequency assignment plan would have to try to use proven, near-term technology to accommodate the sum total of the long-range future requirements of all countries that might possibly want to operate satellite systems . . .
>
> *Freezing of Technology*
>
> A 1977 WARC-type plan would freeze technology which . . . would deny the very increases in orbit-spectrum capacity and decreases in cost required for continued efficient and economical use of these limited resources that Article 33 tells us are essential for equitable access.
>
> *Inhomogeneity*
>
> [. . .]This inhomogeneity is intrinsic to fixed-satellite networks. Such systems must accommodate in an economic and dynamic fashion a wide variety of numbers and types of channels, even within a single national system. This is why fixed-satellite planning is radically difference and much more complex than the broadcasting-satellite planning of 1977 which was based on the assumption of completely homogeneous systems.

Unused Orbit/Spectrum Resources

[...]The interlocking nature of a plan would make it practically impossible for other countries to make use of unused assignments.

Inflexibility

The inflexibility of a detailed plan also makes it difficult if not impossible to modify a plan to provide assignments for new countries, or for new regional groupings of countries wishing to establish a shared system... Thus a plan will tend to institute, *de facto*, a permanent or semi-permanent priority, contrary to the principle stated in Resolution No. Spa 2-1.

For all these reasons, the United States holds that the words "planned" and "planning" in the draft Resolution must be interpreted in a broad and flexible sense. With this interpretation of planning and with the obligation to consider a variety of alternative approaches to the use of the orbit-spectrum resource by space services, the next Space Conferences can lead to the full realization of the objective of equitable access to the resource that the United States has always supported. [61]

2.2. View from India

India, the leading Third World space science power, ran into coordination difficulties attempting to establish its own domestic satellite system. Delegate Srirangan of India explains India's motives in seeking *a priori* allotments in the statement quoted below:

> [W]e decided to embark on a domestic satellite system which is now in implementation. In this process, we had to engage in a very intensive dialogue with various other administrations whose services were likely to be affected by our proposal for a domestic satellite system which provided for both broadcasting and telecommunications services. It was in this process of international dialogue that we encountered considerable problems in insuring a reasonable location in the orbit, and also in insuring appropriate frequency assignments for ourselves. After prolonged dialogue, we have determined the location. But, in that process, in our judgment, we have paid a fairly heavy and severe penalty.
>
> We wrestled with this for two, two-and-a-half, years in our preparatory efforts for the conference, to see what approaches would be possible to insure a reasonable, equitable, timely, and guaranteed access to this limited resource by all countries. And particularly, naturally, we did take into account the interest of the developing countries, because they are the people who seek access much later,

> whose resources are limited and who are in fact not in a position to pay any penalties ... [T]he only practical solution to this problem would be to have an orbit and frequency assignment approach. We found that there was no reasonable acceptable alternative to this dilemma. And in our judgment, the planning approach seemed to present the only acceptable one —whatever be its limitations — because all other approaches seemed to suffer from even greater limitations. [62]

Conclusion

The ITU is facing an issue that threatens the common interest integrity of the organization. The bonds of national self-interest and countries' mutual needs to communicate, which have held the institution together and made it a model of success for other United Nations specialized agencies, can no longer contain the debate on satellite communications' use of the orbit/spectrum resource to technical parameters and engineering designs. Very visible and very political issues are at stake.

Use of information and data-processing technologies depend upon efficient telecommunications networks. Telecommunications has blurred distinctions between domestic and international information flows, and concomitantly, the effects of those flows. The ITU is the international organization most directly involved in attempts to regulate and manage changes due to the infusion of information technologies permeating all aspects of society on both the international and domestic levels. Today, more than ever before, the ITU has an immediate presence in its members' domestic policy-making processes.

This is perhaps the most important change in the ITU's policy-making environment. In the "Information Age," ITU actions have an unmistakable domestic, as well as international, effect in determining economic, political, and military winners and losers. With this in mind, it is apparent why access to the orbit/spectrum resource has politicized the ITU; use of communication satellites facilitates access to information, and thereby, power.

To the extent that maldistributed access to the geostationary orbit —which favors the few space powers — mirrors the larger overall asymmetrical distribution of global power, the ITU is being politicized by technologically powerless states attempting to exploit a slightly more favorable decision-making structure inside the organization. In other words, the LDCs are skillfully using the access-planning issue as a means to exchange HDCs' need for a cooperative spectrum-use regime to obtain concessions telecommunications on development funding, information sovereignty and, more importantly perhaps, to gain a decision-making input in the accelerating militarization of space.

REFERENCES

1. Armando Vargas, "A Challenge for the Third World," *Intermedia*, July-September, 1982, p. 30.
2. The definitive history of the ITU up to 1952 is the work by George A. Codding, Jr., *The International Telecommunication Union: An Experiment in International Cooperation*, (Leiden: E.J. Brill, 1952), and his later work done in conjunction with Anthony M. Rutkowski, *The International Telecommunication Union in a Changing World*, (Dedham: Artech House, 1982), pp. 3-21.
3. See Codding, "The ITU and the Plenipotentiary," in *Intermedia*, (July-September 1982), center insert. A most illuminating description and analysis of the ITU legislative and juridicial processes is found in the book by David M. Leive, *International Telecommunications and International Law: The Regulation of the Radio Spectrum*, (Dobbs Ferry: Oceana Publications, 1970), pp. 29-73.
4. The earth is subdivided into three regions for ITU frequency allocations. Region 1 is all of Europe, Africa, and the Soviet Union, Region 2 is North and South America, Greenland, and Hawaii, Region 3 is Asia (except for the USSR), Australia, and Oceania.
5. The three delineations of spectrum rights apportioning were derived from the insightful comments made by Professor Carl Christol at a panel on the politics of outer space, Western Political Science Association Conference, San Diego, March 28, 1982. There are still, however, problems in applying the definitions of allotments and assignments in the IFRB's coordiations procedures. See IFRB, Circular-Letter No. 600, ITU, 10 December 1984, Geneva, pp. 22-31.
6. See, ITU, Final Acts, World Administrative Radio Conference, 1971, 1977, and 1979.
7. ITU, CCIR, IWP 4/1, Report on Efficient Use of the Geostationary Orbit, 1982.
8. For example, the fact that there are "only" three color television transmission and reception standards is often used as an example of CCIR effectiveness. The United States, French, and West German systems for broadcasting and receiving color television are the three worldwide standards. Most experts contend that the global distribution of television programming and equipment would be greatly restricted if each country used its own television standards, making its programming incompatible with other systems.
9. Anthony M. Rutkowski, "The International Telecommunications Union and the United States," *Telecommunications*, October 1983, pp. 35-43, at 36. The United States share of the ITU budget is about seven percent, with the total HDC share amounting to approximately 55 percent.

10. See, Codding and Rutkowski, pp. 113-114.
11. Interview with Dr. Bruce Lusignan, Palo Alto, California, February 7, 1984.
12. Bruce Lusignan, "Efficient Satellite Services in Developing Countries: Availability of Inexpensive Earth Stations," Paper presented at GLOBECOM '83, San Diego, California, November 29, 1983.
13. Codding, 1982, pp. 102-103.
14. The present Director of the CCIR, Richard Kirby, nominated by the US, was re-elected at the 1982 CCIR Plenary Assembly which was held before the November 1982 Plenipotentiary Conference. He will serve out his normal term of office. However, the United States "deplores the politicization of the ITU" which this change represents, commented David Macuk, Telecommunications Attache, US Mission, Geneva, in an interview on December 8, 1982. In Macuk's view, this change increases the likelihood that the US will "lose" this influential position at the next Plenipotentiary Conference, another factor prompting "a United States reassessment of its participation in the ITU." The directorship of the CCIR was considered "American property," in the same way directorship of the CCITT was considered "French," and that of the IFRB as a "Russian" post.
15. ITU, IFRB, "Frequency Management and the Use of the Radio Frequency Spectrum and of the Geostationary Satellite Orbit," 1979, p. 5; see, Codding and Rutkowski, 1982, p. 122.
16. Quoted from F.M. Negro and J.N. Novillo-Fertrell y Paredes, "The International Telecommunications Convention from Madrid (1932) to Nairobi (1982): half a century in the life of the Union," in *Telecommunication Journal* 49(December 1982):816.
17. ITU, IFRB, "Frequency Management and the Use of the Radio Frequency Spectrum and of the Geostationary Satellite Orbit, Geneva, 1979, p. 5; as quoted in Codding and Rutkowski, 1982, p. 123.
18. Readers wishing a more comprehensive treatment of IFRB procedures are encouraged to refer to Codding and Rutkowski, 1982, p. 124; and Leive, 1970, especially Chapters 2, 3, and 4.
19. E.O. Ducharme, R.R. Bowen, and M.J.R. Irwin, "The Genesis of the 1985-87 ITU World Administrative Radio Conference on the Use of the Geostationary Satellite Orbit and the Planning of Space Services Utilizing It," *Annals of Air and Space Law* 8(1982):269-270.
20. DuCharme, *et al.*, p. 271.
21. Fabrio Galante, "An Overview of the Institutional and Regulatory Aspects and Their Impact on System Design," paper presented at the International Astronautical Federation Congress, October 1983, Budapest, Hungary.
22. *Ibid.*

23. Personal interview with Edward Jacobs at the IFRB POE meeting in Geneva, December 8, 1982.
24. ITU, Final Acts, WARC 1979, Article 11, para. 1047, 1066. Furthermore, under the Regulations administrations are required to notify the IFRB only when it is suspected that the satellite system may cause *international* interference.
25. ITU, Final Acts 1959 WARC, Recommendation 36.
26. See, generally, ITU, *From Semaphore to Satellite*, (Geneva: ITU 1965), pp. 300-301.
27. EARC, Final Acts, 1963.
28. See, generally, ITU, Final Acts for the World Administrative Radio Conference for Space Telecommunications, July 17, 1971, U.S.T. 1527, 1686; T.I.A.S. No. 7435.
29. ITU, Final Acts, WARC-ST, 1971, Res. Spa 2-1.
30. Christol, 1982, p. 558.
31. Interview with international legal scholar Eilene Galloway, December 2, 1982, Washington, D.C. Ms. Galloway interpreted the provision to allow replacement of a satellite that expired prematurely, i.e., before the average seven-year engineering lifetime had passed. This was the case, for example, when RCA Satcom III was lost shortly after the firing of its apogee motor in 1979. A relatively short time later, RCA launched Satcom III-R to the same orbital slot.
32. Martin A. Rothblatt, "ITU Regulation of Satellite Communication," in *Stanford Journal of International Law*, 18 (Spring 1982):1-25, p. 9.
33. *Ibid.*
34. Christol, 1982, p. 561.
35. *Ibid.*, p. 560.
36. ITU, International Telecommunication Convention 1973, Article 33.
37. ITU, International Telecommunication Convention 1973, Article 10(3)b-d; as emphasized by Rothblatt, 1982, p. 10, fn. 44.
38. More specifically, these frequencies were 11.7-12.2 GHz in Regions 2 and 3, and 11.7-12.5 GHz in Region 1.
39. Rothblatt, 1982, p. 11.
40. *Ibid.*
41. Denis-Lempereur, Jacqueline, "La vraie reforme nous vient du ciel." *Science et Vie*, May 1982, pp. 72-201.
42. Interview with M. Sant of the IFRB, Pacific Telecommunications Conference, Honolulu, January 15, 1985.
43. See, in general, *Intermedia* (August 1982), and Guy M. Stephens, "Will DBS Bridge Its Troubled Waters?" *Satellite Communications*, February 1985, pp. 34-35.
44. Codding, 1982, p. 48.

45. ITU, Final Acts WARC-ST, Article VII, Section 428A, (#2674 of the 1979 Radio Regulations), and #2612 of the 1979 Radio Regulations; these were emphasized by Secretary-General Butler in his monograph, "The International Telecommunications Law of Satellite Broadcasting," September 1984. See also, Carl Q. Christol, "Telecommunications, Outer Space, and the New World Information Order," *Syracuse Journal of International Law and Commerce* 8:2(1982):343-364, at 360; and Theodore M. Hagelin, "Prior Consent or the Free Flow of Information over International Satellite Radio and Television: A comparison and Critique of U.S. Domestic and International Broadcast Policy," *Syracuse Journal of International Law and Commerce* 8:2(1981): 265-320.
46. Interview with Secretary-General Butler, December 10, 1982, Geneva. See, Larry F. Martinez, "Butler Ushers the ITU into the 1980s," *Satellite Communications*, March 1983, p. 43.
47. See E.O. Ducharme, R.R. Bowen, and J.J.R. Irwin, "The Genesis of the 1985-87 ITU World Administrative Radio Conference on the Use of the Geostationary Satellite Orbit and the Planning of Space Services Utilizing It," *Annals of Air and Space Law*, 8(1982), pp. 261-281. The authors write that at the 1971 WARC-ST:

    France, Argentina, and Brazil at the outset, were adamant that satellite broadcasting in the 12 GHz frequency band should not be implemented prior to the formulation of plans. The highly political overtones and the social and cultural impact of direct broadcasting satellites (DBS) together with the question of radiation spillover (intentional or otherwise) in the territory of other countries, were factors which prompted the planning of DBS. It was believed that the *a priori* planning and assignment of radio channels, orbit positions and associated coverage areas to individual countries would significantly alleviate the spillover problems. (p.267)

48. See Chapter 3's discussion on appropriation of the outer space environment.
49. Rothblatt, 1982, p. 12.
50. A "General" WARC differs from other WARCs, such as WARC-ST and WARC-BS, in that it has the task of reviewing and revising the Radio Regulations for all services in all allocated frequency bands.
51. Brian Segal, "International Negotiations on Telecommunications," in *Intermedia*, November 1980, pp. 21-31, at 21.
52. Glenn O. Robinson, "Regulating International Airwaves: the 1979 WARC," *Virginia Journal of International Law*, 21(1980):1-50, at 12.

53. See, Bert Cowlan, "UNISPACE '82: A Retrospective," Paper presented to Office of Technology Assessment during Technical Memorandum Meeting, December 2, 1982, Washington, D.C.
54. David Honig, "Lessons for the 1999 WARC," *Journal of Communication*, 30(Spring 1980):48-58; Honig cites, M. Porat, "Communications Policy in an Information Society," in *Communications for Tomorrow: Policy Perspective for the 1980s*, (New York: Praeger, 1978); and Heather Hudson, "WARC '79: Development Communications Strategies — A Report to USAID," Academy for Educational Development, Washington, D.C., 1979.
55. Quoted in Honig, p. 50.
56. Robinson, p. 13.
57. Segal, p. 24.
58. For an extensive review of the WARC-79 Final Acts, see the excellent report written by the US Congress, Office of Technology Assessment, *Radiofrequency Use and Management: Impacts from the World Administrative Radio conference of 1979*, January 1982. The Final Acts fill a volume of 1,100 pages containing the revised Radio Regulations, the WARC-BS Final Acts, Annexes, Resolutions, Recommendations, Reservations, and various protocols submitted by states.
59. ITU, Final Acts WARC-79, Resolution BP.
60. A.M. Rutkowski, "Six Ad-Hoc Two," *Satellite Communications*, March 1980, p. 25. Rutkowski traces the development of the Resolution during debates held in the Working Group. As Rutkowski states,

    [T]he actions it took appear to have markedly changed the course of space communications law . . . [I]t was the first and only significant global interchange of views on mechanisms for vesting rights for the geostationary orbit for telecommunications (p. 22).

61. T.V. Srirangan's comments are quoted from A. Rutkowski, "Six Ad-Hoc Two: The Third World Speaks Its Mind," *Telecommunications*, March 1980, p. 23.

CHAPTER 5

THE HIDDEN AGENDA REVEALED

Hey, what these guys are talking about could affect my business! [1]

- *W.F. Wright,*
Vice-President, NASA Systems, Lockheed Corporation

We are not just dealing with technical aspects, we're talking about security aspects. The use of the geostationary orbit is linked to the questions of the militarization of space and the consensus here is that we are opposed to this ... Developing countries must defend themselves. [2]

- *Hector Charry-Samper,*
Colombian Delegate to UNISPACE '82

The country with sole possession of space lasers would have "the longest 'big stick' in history," nothing less than "the capability for unilateral control of outer space and consequent domination of the earth." [3]

- *Dr. John D.G. Rather,*
Vice-President, Kaman Aerospace Corporation,
quoted in New York Times

Space Problems in the Hofburg Palace

The Second United Nations Conference on the Exploration and Peaceful Uses of Outer Space, or "UNISPACE '82," was a high-level diplomatic discussion of mankind's use and management of the outer space environment. UNISPACE '82, attended by more than 1000 official delegates from 94 countries and over 30 international organizations, was held August 9-21, 1982, in the elegant Hofburg Palace in Vienna, Austria.

UNISPACE '82 was a much more politicized conference than its 1968 predecessor. During the 14-year interim, outer space had become an area of global commercial and military competition as countries sought to develop or acquire space technology and access to the space environment for a growing number of political, economic, and strategic objectives. By the 1980s, both the number and diversity of states with active space programs had grown from the two superpowers to include Japan, Western Europe, the Soviet Bloc, and the Third World spacepowers — India, Indonesia, and China. In a significantly visible way, UNISPACE '82 demonstrated that a majority of countries now see their national interests at stake in how a global regime for the outer space environment is established.

UNISPACE '82, ostensibly technical in scope and purpose, was in fact a political event from start to finish. It was held at the behest of Third World countries, concerned that they were about to be left further and further behind as the spacepowers give free rein to their galloping technologies. But one issue in particular dominated the UNISPACE '82 discussions — space militarization. In short, the Third World intent behind convening UNISPACE '82 was to place outer space issues in the global spotlight and to focus international attention onto the intensifying effort by the superpowers to extend their strategic competition into the space environment. In this way, UNISPACE '82 revealed the hidden agenda politicizing the supposedly technical negotiations regarding allocations and management of that very favorable area of outer space — the geostationary orbit.

The Conferences

The administrative purpose of UNISPACE '82 was to approve the final draft of "Report on Outer Space," submitted by COPUOS, that, following conference additions, deletions, and amendments, would be presented to the UN General Assembly. The report would represent world consensus on steps the General Assembly should take to "preserve outer space for purely peaceful purposes" and to promote a wider distribution of the benefits derived from space technology and exploration [4].

The Draft Report arrived in Vienna with a consensus agreement on all but 15 of the 435 paragraphs. It was divided into three sections which would be reviewed by three separate committees: (1) the current status of space exploration and use, (2) applications of space technology and resources, and (3) methods for promoting international cooperation and distribution of the benefits from space exploration and utilization.

Following the COPUOS precedent, decisions approving paragraphs were reached through consensus. Voiced disapproval by one state's delegate was sufficient to block inclusion of a paragraph. Although only 15 paragraphs arrived under contention from the original drafting committee, discussions were long and exhausting in the sticky Viennese summer heat. Many involved "trivial" details on already accepted paragraphs. Some experienced observers of UN space-related conferences attribute this "pickiness" to the need of some delegates from financially-pressed foreign ministries to produce some form of tangible results from their expensive junket to Vienna. Adding it all up, UNISPACE '82's price tag came to more than $30 million, a sum greater than many of the development programs proposed at the conference but rejected on the basis of cost.

UNISPACE '82 itself was only one of three space conferences taking place simultaneously in Vienna. As an "alternative" to the official UN conference, some 20 non-governmental organizations (NGOs) assembled experts to give talks and lead discussions on an extensive array of outer space topics in the Messepalast, one kilometer from the Hofburg Palace. Anticipating a tepid official conference in which potentially divisive issues such as space militarization would be blocked by the superpowers, the NGOs intended to provide the forum where these questions could be presented. It quickly became apparent, however, that the superpowers were unable to contain or control the direction of the discussions and UNISPACE '82 became a scene of intense negotiations, bargaining, and not a little drama.

The NGO conferences became an adjunct to the third area of interest —the exhibitions of space technology. Twenty-five countries and four international organizations, including INTELSAT, INMARSAT, and ESA (European Space Agency), sponsored public exhibits of their national space programs. The exhibitions were also the sites of press conferences and served as integral parts of the states' and organizations' overall lobbying and public relations efforts. Outside on the lawns between the Hofburg Palace and Messepalast Exhibition Hall one could see some 30 satellite dish antennas, pointing skyward as silent reminders of the centrality of the geostationary orbit to the discussions inside.

Rather symbolically, the exhibits of the United States and Soviet Union stared at each other across a courtyard, each depicting its own space programs and achievements in its own distinctive national style. The differences were striking. Upon entering the Soviet exhibition, the first displays to be seen were four bronze busts of dead cosmonauts. In contrast, the United States exhibit made maximum use of videotapes showing dramatic Space Shuttle launches and landings. Each superpower made maximum use of its astronauts and cosmonauts in a public relations battle to show its space program to be the most beneficial to the Third World [5].

The Orbit/Spectrum Debates

Alongside space militarization, the geostationary orbit was the most contentious issue discussed at UNISPACE '82. UN Secretary-General Perez de Cuellar, speaking at the opening ceremony, commented that "because the geostationary satellites see one-third of the earth, their use necessitates cooperation and at the same time confrontation between users." [6] The Secretary-General's statement proved to be an accurate prediction of the negotiations to come. In many ways, though, the orbit/spectrum debates at UNISPACE '82 were a continuation of the ITU discussions in WARC-79's subcommittee Six Ad-hoc Two (discussed in Chapter 4) was attended by many of the same delegates [7].

the access and use dimensions of the orbit/spectrum controversy: (1) resource scarcity and planning, and (2) use for "the benefit of all nations." We will now examine the committees' deliberations over the orbit/spectrum resource and the ways in which they serve as indicators of the motives behind the increasing politicization of outer space and satellite communications.

1. Resource Scarcity and Planning

The First Committee, dealing with the current state of space exploration and use, adopted the following assessment of resource scarcity and capacity:

Paragraph 64

> Most communication satellites are in the geostationary orbit. Since these satellites operate within limited designated frequency bands and therefore need to be sufficiently separated to avoid interference with one another and also possible collisions, there is a limit to the total number of satellites that can operate in the geostationary orbit in the different frequency bands and also in any given frequency band. There is real concern that some parts of the orbit are approaching saturation in certain frequency bands. Technological advances are underway, however, which will probably permit among other changes, the closer spacing of satellites and their satisfactory coexistence. It is therefore imperative that studies and research to achieve this objective be intensified, including a closer examination of techno-economic implications, particularly for developing countries in order to ensure the most effective utilization of this orbit in the interest of all countries. Besides those efforts, and bearing in mind that the geostationary orbit is a limited natural resource, thus it is imperative that its use be properly and justly regulated. [8]

These "techno-economic implications" dominated the discussions concerning access to the orbit/spectrum resource. An Indonesian delegate stated the Third World perspective of the issue by saying,

> A non-technological solution to the geostationary orbit crowding is needed by LDCs; it is a limited natural resource and the need to regulate is imperative. The problem is that costs go up with changes in technology. [9]

The United States delegation, meanwhile, was busily engaged in its effort to emphasize how a rapidly evolving technology was steadily enlarging the telecommunications capacity of an orbital slot. Available at the document centers was a fold-out pamphlet in English, French, and Spanish entitled "The Keys to Communications Growth." Its findings conclude that:

> In the foreseeable as well as the distant future technological progress will assure equitable access to the geostationary orbit at affordable cost for users of the radio spectrum. [10]

Curiously, the pamphlet was distributed anonymously by the United States so that its views could be presented as "scientific fact" and not just those of a particular country, in this case the major protagonist of flexible orbital management [11].

Costs link scarcity to planning. While the US delegates tried to down play scarcity arguments expounded by Third World delegates (a difficult position to explain since the FCC had already recognized geostationary congestion over the US), they were nonetheless cognizant of the inherent cost factor in any form of resource conservation. American policy flexibility, however, was emphasized during a "Poster Session" led by United States delegate Harold G. Kimball, Director of the communications and Data Systems Division of NASA, who responded to this author's questions regarding the cost of orbital saturation by saying:

> This is a delicate point. Even though the per circuit hosts are down, to leap into the more efficient high technology is more expensive. However, countries that cannot buy the technology can rent it through INTELSAT. The United States does not make the argument that all planning is undesirable. [12]

LDCs, though, were facing numerous committee defeats to pass paragraphs that they felt would facilitate or reduce their access costs. For example, one such paragraph [13] called for the setting aside of the less problematical lower frequency bands for developing country use while the HDCs would be encouraged or required to move their telecommunications upward into the rain-attenuated Ku- and Ka- frequency bands. Outside the committees, Third World countries adopted a statement emphasizing their commitment to reducing access costs through detailed planning of the orbit.

Group 77 Position Paper (excerpt)

> . . . Group 77 is of the firm view that the present regulatory mechanism for assigning orbit positions and radio spectrum does not ensure equitable access to this resource, and that the developing countries are particularly at a disadvantage. Group 77 is, therefore, of the view that a change to this mechanism is called for. Group 77 notes that WARC-79 of the ITU examined this problem in detail and decided to convene a Special Conference and Regional Conferences "to guarantee in practice for all countries equitable access" to the said resource and to agree on appropriate planning and other approaches to fulfill this objective. Group 77 considers that the principle of guaranteed and equitable access should be the essence of any new regulatory mechanism and should take into account the particular needs of the developing countries including those of the equatorial countries. [14]

By incorporating the key word, "planning," the statement represents the Group 77's seemingly clear endorsement of detailed planning to achieve "equitable access." On the other hand, it also illuminates divisional strife within the LDC bloc. Inclusion of the word "needs" instead of "rights" indicates that the position paper attempted to skirt dissension among G77 members who dispute the equatorial countries' claim of geostationary sovereignty [15].

"Needs" can refer to both legal and technical requisites for orbit/spectrum access. Arriving at a consensus concerning what constituted a broad definition of "equitable access" was extremely difficult to achieve given the lack of precise boundaries marking the beginning of the outer space regime and the legal status of the geostationary orbit. The debate revolved around paragraphs 281 and 284.

Paragraphs 281 and 284 contained references to the equatorial countries' claims of sovereignty over "their" portions of the orbit and that the orbit should be governed by a *sui generis* regime — in other words, arrangements separate from the Outer Space Treaty. In a dramatic late-night session at the end of the two-week conference, a drawn-out debate on the paragraphs was resolved through the adoption of compromise wording mentioning but not supporting the Bogota Declaration claims. Significantly, consensus was also reached on the scarcity and planning provisions in the paragraphs, supposedly the most divisive of the orbit/spectrum issues.

2. For the "Benefit of All Nations"

Underlying the technical and legal debates lay the power dimension behind UNISPACE '82: Does use of outer space benefit all countries or does it serve to widen the gap between space and non-space powers? National sovereignty concerns are at the core of the factors politicizing outer space and the orbit/spectrum resource. Direct broadcast and remote sensing satellites pose the questions of whether states have an internationally-recognized right to control the flow of information over their national boundaries, even if that information is collected or disseminated by other states in pursuit of their own sovereign prerogatives. This issue of "prior consent" remains an unresolved point of international law to this day.

In a strategic sense, the conference marked a turning point in the evolution of the emerging regime for space. Space, and in particular the geostationary orbit, had become integral components of the global balance of power. UNISPACE '82 was politicized by the perception held by many Third World countries that space technology determines the worldwide distribution of power. For these countries, the militarization of space by the superpowers represents a massive shift in global power to the superpowers.

These countries saw the increasing numbers of military satellites and space weapon development programs as antagonistic to their own strategic objectives of redressing the imbalance of power between themselves and the superpowers. Consequently, they used the conference as a forum in which they could express their views concerning the rules of the road in space, i.e., what a state can and cannot legally do. Third World countries being technically powerless to prevent this impending shift in power from taking place, used other issues (e.g., the orbit/spectrum access issues) as fora for expounding their concerns to gain bargaining leverage so as to make space militarization as politically costly to the superpowers as possible.

This tactic is seen most clearly in a statement made by the Colombian delegate, Hector Charry-Samper, in support of the equatorial countries' claims to geostationary sovereignty. Significantly, Charry-Samper does not emphasize resource saturation or the cost of technical access, but instead, the strategic motivations behind the LDCs' demands:

> There is now a *de facto* occupation of the orbit. The developing countries are trying to defend themselves because they find themselves in a state of disequilibrium vis-a-vis those who are appropriating the orbit, not in the interests of humanity. They are not making an attempt to preserve outer space for the use of the whole of humanity, as has been declared. *We are not just dealing with technical aspects, we're talking about security aspects. The use of the geostationary orbit is linked to the question of the militarization of space and the consensus here is that we are opposed to this . . . Developing countries must defend themselves.* [16]

Colombia was not alone in the position it took. The G77 Position Paper also mentioned space militarization as:

> [a]s a matter of great concern . . . Group 77 countries further reiterate that militarization of space is detrimental to the entire humanity [sic] and hence extension of [the] arms race to Outer Space . . . that is the common heritage of mankind, should not be permitted. The Group 77 considers necessary the adoption of a legal instrument that definitely bans the emplacement of weapons in outer space and verifiable controls and guarantees [sic]. [17]

In our discussion of the orbit/spectrum regime, we found that legal scholars consider use of the space environment consistent with the requirements and conditions stipulated by the Outer Space Treaty when carried out under the supervision of a state for the benefit of all countries. However, while Article 4 of the Treaty prohibits the orbiting of "weapons of mass destruction," it is silent regarding the legality of space satellites functioning as integral parts of nuclear weapon systems or in terms of other types of space-based weapon systems such as orbiting laser battlestations.

What, then, is the negotiating link between issues pertaining to communication satellites and the orbit/spectrum resource, and the much higher-level strategic confrontation between the superpowers over space militarization?

Third World countries, cognizant of the tremendous advantages offered by satellite communications and other space technologies for their own development goals, see space weapon technologies pushing the superpowers a quantum leap ahead in terms of their strategic superiority vis-a-vis the rest of the world. Space weapons accomplish two superpower goals.

First, in the view of an increasing number of Third World countries, the superpowers' weapons race into space consolidates their strategic superiority by stabilizing the nuclear parity between themselves and their allies. The Reagan Administration's Strategic Defense Initiative (SDI), while shown by many experts to be unrealistic in its stated objective to protect civilian centers from Soviet ICBM attack, would conceivably function as an effective shield protecting US land-based ICBMs. In this way, the SDI can protect the credibility of the American deterrence and maintain mutual assured destruction as a viable strategic doctrine [18].

Secondly, these weapons offer an added power benefit to the superpowers. In most scenarios, many of the weapons would be placed in earth orbit which would carry them over most or all of the Third World countries. For these countries, the long-term trend is clear: the spacepowers will continue to develop and deploy space weapons that can be legally flown over all countries under the free access guarantees of the Outer Space Treaty.

To Third World countries, these ostensibly "defensive" weapons have many "offensive" attributes which constitute a continual superpower presence and an ability to strike air and ground targets with impunity from the safety of space. As a *New York Times* article stated:

> Some experts believe that such lasers could provide an awesome instrument for surgically precise attacks against "soft" targets ... anywhere on the earth's surface, in the air, or in space from ranges of thousands of miles with no collateral damage to adjacent civilian populations ... Other experts have suggested that fires might be started in grain fields and storage bins, thus starving a country into submission, or that flammable structures might be torched the length and breadth of a country to spread havoc ... [19]

According to Dr. John Rather, Vice-President of Kaman Aerospace Corporation, possession of space-based weapon systems amount to "the capability for unilateral control of outer space and consequent domination of the earth." [20]

The Third World strategy is also clear. Realizing that they are technologically powerless to control or even cool the heating arms race in space, the LDCs are "socializing" the conflict into other decision-making arenas, notable the ITU, where they enjoy voting majorities and where the orbit/spectrum resource offers unique bargaining advantages.

First, the LDCs can exploit HDC vulnerabilities and their needs for a cooperative-use regime for the orbit/spectrum resource. The Bogota Declaration and the call for detailed planning of the geostationary orbit and associated frequency bands have effectively focused world attention on the question of how outer space should be used and managed. Spectrum coercion and jamming are tactics that force the superpowers to listen to the Third World's concerns.

Secondly, the dependency of HDC multinational corporations and economic systems on the free, unfettered flow of business and administrative information over national boundaries provide the LDCs with yet another policy fulcrum with which to gain bargaining leverage against the superpowers and their allies. The militarized and politicized geostationary orbit as a communications resource is the major bargaining link between issues of data protectionism (or in terms of communications satellites, "prior consent") and space militarization. Proposals are already heard in COPUOS to ban the military use of the geostationary orbit. In retaliation for space militarization, Third World countries may make it increasingly difficult or expensive for HDC multinationals to interconnect their widely-flung manufacturing and administrative facilities through the erection of technical or legal data barriers.

Closing Observations

In the final technical analysis, the much discussed scarcity of the orbit/spectrum resource is more myth than math. By 1988, high-capacity undersea optical fiber cables will take much of the pressure off satellite trunks and much of the steam out of scarcity arguments at ITU conferences. Scarce, though, remain the financial resources necessary to establish a satellite system, and perhaps even more importantly, the requisite terrestrial telecommunication infrastructures. Allotments of space resources alone will not enable debt-ridden countries to establish the satellite and other telecommunications infrastructures they need to achieve a modicum of information competence.

If the unstated goal of the Bogota Declaration countries and the proponents of detailed orbit/spectrum planning is to effect some type of "geostationary" rent from HDC satellite operations to pay for LDC networks, the rationale collapses from a weakening technological basis. In addition to fiber optics, an increasing number of common-user networks, improved modulation techniques, and higher satellite power levels all mean greater telecommunications capacities from orbital slots available to more users at less cost. For these reasons, the argument that detailed planning would make satellite networks more affordable to LDCs fails to adequately explain the growing politicization of the geostationary orbit and radio spectrum resources.

At first glance, the increasing importance of information technologies and satellite networks as factors of global and national power would seem to explain the politicization of the resources. But if the resources are not "scarce" in any long-range technological sense that would prevent access, why are they increasingly controversial issues?

The reason can be found in the perceptual linkage between use and access. The crucial perception is that these outer space resources as they are now and will be used widen the gap between North and South. While the present ITU procedures for allocating and managing access to the resources may only indirectly affect the power gap, the ITU (and COPUOS) offers the LDCs an advantageous bargaining forum. It is here that the LDCs are attempting to exploit HDC vulnerabilities to spectrum coercion and dependency on the free flow of information in order to gain a voice in the larger strategic debate over space militarization. As long as use and access are perceived by many LDCs as linked, the politicization of the orbit/spectrum resource and satellite communications will continue, and in some new directions.

REFERENCES

1. Informal comment to the author while observing UNISPACE '82 discussions, August 13, 1982, Vienna.
2. Statement given by Colombian Delegate Hector Charry-Samper, August 19, 1982, Vienna.
3. Quote of Dr. John Rather from article by Philip M. Boffey, "Dark Side of 'Star Wars': System Could Also Attack," *New York Times,* March 6, 1985.
4. Opening statement by UNISPACE '82 President Yash Pal, Vienna, August 7, 1982.
5. Office of Technology Assessment, "UNISPACE '82: A context for International Cooperation and Competition," Technical Memorandum, March 1983. The OTA report was critical of the US approach to the conference in that the US "has allowed itself to become isolated on the military and DBS issues, . . . ", p. 8.
6. Statement by UN Secretary-General Perez de Cuellar, Vienna, August 9, 1982.
7. Bert Cowlan and Lee Love, "UNISPACE '82: A Retrospective," submitted to the US Congress, Office of Technology Assessment, November 9, 1982, pp. 161-205.
8. UNISPACE '82, Report of the Second United Nations Conference on the Exploration and Peaceful Uses of Outer Space, A/CONF.101/10, 1982. This paragraph appears as number 66 in the Final Report.
9. Author's notes, Vienna, August 12, 1982.

10. Unidentified author, "The Keys to Communications Growth," distributed at UNISPACE '82.
11. Interview with Donald Jansky, NTIA, Vienna, August 17, 1982. Author's transcripts.
12. Author's notes, Vienna, August 16, 1982.
13. Paragraph 253 of the Draft Report
14. UNISPACE '82, A/Conf.101/5, Vienna, August 13, 1982.
15. Interview with David Small, US State Department, Vienna, August 14, 1982.
16. Colombian delegate Hector Charry-Samper during debate on paragraphs 281 and 284, Vienna, August 19, 1982. Author's transcripts.
17. UNISPACE '82, "Declaration of the Group of 77," A/Conf.101/5, Vienna, August 13, 1982.
18. Phillip M. Boffey, "'Star Wars' and Mankind: Consequences for Future," *New York Times,* March 7, 1985.
19. Phillip M. Boffey, "Dark Side of 'Star Wars': System Could Also Attack," *New York Times,* March 6, 1985, p. A-24.
20. *Ibid.*

APPENDIX

A THEORETICAL PERSPECTIVE FOR COOPERATION*

Introduction

The challenge facing the ITU, COPUOS, and other international policy-making bodies concerned with outer space, is how to formulate rules and rights that ensure efficient and equitable resource allocations and management, which, at the same time, promote compliance by respecting global configurations of power. Regime rules and rights are not self-generating; as international law, they mirror changing configurations of (1) technology, (2) political power, (3) economic interests, and (4) military might in the international system. This section will show how the *collective goods* model of micro-economic analysis mirrors the configurations of states' interests and power within the international system. I will argue that the collective goods model usefully illuminates incentives (or disincentives) for governmental participation and contribution towards maintaining a cooperative-use regime for the orbit/spectrum resource and satellite communications.

From the collective goods perspective, the politicization of the rules and rights for the orbit/spectrum resource stems from a growing incongruence between rules for the *nonexcludable* spectrum resource, and the proliferation of national and regional satellite systems utilizing increasingly scarce orbital slots with *excludable* benefits.

The Setting

The problem is this: satellite and information technologies have created a new global commons resource — the geostationary orbit — and as yet the international system has been unable to integrate it into the pre-existing ITU cooperative-use regime built around the spectrum resource. The model argues that special characteristics of the spectrum resource compels international cooperation between users, while this is not the case with the geostationary orbit.

There seems to be a growing contradiction between the *legal reality* of free and open access to the space environment, as stipulated by the Outer Space Treaty, and the *political* and *technical* realities perceived as the growing scarcity of favorable portions of the geostationary orbit in the most desirable frequency bands (especially the C-Band).

Since the earliest days of radiocommunications, countries have pursued their individual self-interests through joint collaborative arrangements in

*An earlier version of this appendix appeared in the AIAA *Proceedings of the 26th Colloquium on the Law of Outer Space*, 1983.

their use of the spectrum resource. Today, the ITU represents the institutionalization of rules and rights constituting the cooperative-use regime for the spectrum resource, and, with the advent of satellite communications, the geostationary orbit resource as well. However, with the emergence of satellite communications, the ITU finds itself immersed in a political controversy that threatens to disintegrate the technical bonds which have held the organization together for over a century. How can this be explained by the collective goods model?

1. The International System

The ITU functions in a global system of sovereign states, each attempting to maximize its power and security while minimizing its dependency on, or vulnerability to, factors outside its control. The primary goal of any state is its security and survival in an anarchic and threatening international system. With no "higher power" to enforce an international law guaranteeing national sovereignty, countries seek to reduce their vulnerability to actions of other states. There is, therefore, an inherent tension between a state's perceived need to cooperate and its instinctual guarding of sovereign prerogatives. In other words, an international cooperative relationship is accomplished at a vulnerability "price." Consequently, states will participate in the ITU regime if the perceived benefits outweigh imposed dependency costs.

2. The Dichotomy Between the Orbit and Spectrum Resources

Satellite communications have changed many of the perceived incentives for participation in the ITU regime, as is evidenced in the growing politicization of ITU Conferences. This is partly due to a fundamental incongruency between the orbit and spectrum resources and the way it can be exploited for access rights and other goals. While the spectrum is universally accessible, the geostationary orbit is not. Nations are excluded from the geostationary orbit by lack of financial, technical, and managerial means. Asymmetrical access to the orbit/spectrum resource translates to an unequal distribution of other resources crucial to political, economic, and military power.

Among countries, the unequal distribution of the technological and financial means required for access to the orbit/spectrum resource magnifies the power advantages of those countries with national satellite networks. Given the importance attributed to telecommunications in general and satellite communications in particular, the maldistribution of *access* to the orbit/spectrum resource is perceived to exacerbate the maldistribution of *power* in the international system.

3. The Problem of Promoting Cooperative Use of Resources

While it is recognized that optimal provision of satellite communication channels is accomplished through cooperative, rather than "competitive" use of the orbit/spectrum resource, the power implications of maldistributed access threaten to polarize the ITU, leading to regime disintegration and a situation of orbit/spectrum anarchy. In order to maintain a cooperative-use regime, the international system must formulate rules and rights that ensure equitable as well as efficient allocations and management of the orbit/spectrum resource. International law can promote cooperation if it accurately mirrors incentives for compliance.

The Model

Analytical models explaining economic relations between individuals in a free market situation offer insights to the behavior of sovereign states in an anarchic international system [1]. Transposing the premises of the micro-economics perspective, we can think of the international system as a form of market, where the actions of countries are analogous those of individuals as they produce and purchase goods which they believe will maximize their interests. Our model is based on the following assumption: states are utility maximizers; their chief criterion of national interest is power, the opposite of which is dependency or vulnerability to factors outside its control. Under what circumstances and in response to which incentives will they "buy into" a regime?

1. The Production of Regimes

In an economy, the production or consumption of certain goods involves externalities or spillover effects, where the benefits or costs are not limited to only the participants in the transaction. This so-called "market failure" skews the price and demand signals resulting in a less than optimal allocation and utilization of resources. A regulatory agency mitigates the effects of market failure by politically enlarging the transaction to include the actual set of producers and consumers, taxing and compensating those affected so as to accurately portray supply and demand signals.

Similarly, use or misuse of the spectrum resource incurs spillover effects, where non-participants in a particular communication mode suffer interference. The radio spectrum's intrinsic universal accessibility and ubiquity characteristically lead to market failure (less than optimal utilization) unless all affected participants are brought into the transaction.

The ITU brings spectrum users into a common transaction. Although it cannot tax and compensate for spectrum externalities, the ITU does distribute benefits in the form of rights to interference-free spectrum (and orbital slots) while exacting payments in terms of countries' cooperation and compliance with the regime rules. The effects of market failure are diminished and optimal utilization is encouraged through the ITU's regulatory capacity to assess costs and allocate the benefits of efficient spectrum use.

States, as rational consumers, "buy" or participate in the regime if the benefits received — in the form of interference-free spectrum and orbital slots for satellite networks — outweigh the dependency costs imposed by participation and cooperation in the ITU. The level at which an international regime is "produced" is established by the value at which the factors of supply and demand coincide.

The regime is "supplied" by a resource-imposed distribution of power among users in the form of spectrum externalities (i.e., interference), where any user can "jam" another. The regime is "demanded" by states in order to receive regime-provided goods of interference-free spectrum and orbital slots for satellite networks, which, due to their utilization of a common property resource, also involve externalities. Goods or resources with externalities are often classified as "collective goods." [2]

2. *Collective Goods*

Purely collective (or "public") goods are classified by two characteristics: they are indivisible and nonexcludable. Collective goods exhibit varying "mixes" of nonexcludability and indivisibility. These are determined by the nature of the good itself, or by the system in which it is produced.

2.1. *Indivisibility*

> "Indivisibility" or "jointness of supply" means that once a good is produced it is equally available to everyone. Extension of the supply to one individual facilitates its extension to all. And, up to a point, its extension to one additional individual does not imply a corresponding reduction in the quantity of the good available to others. [3]

For example, the audience receiving a television broadcast is sharing an indivisible good; additional viewers do not detract from the quality of the good, nor do they impose additional costs upon the producer or other consumers.

2.2. *Free Access or Nonexcludability*

Collective goods are also characterized as freely accessible or nonexcludable. There are two aspects of nonexcludability. The first concerns the ability of the provider to exclude:

> Instead of being able to limit the utilization of it to those participating [and paying] in its production, a state [or regime as producer] may confront "impossibility of exclusion"... [where] it may not be possible to exclude others from sharing or to charge them the full cost of sharing the benefits of the good. [4]

The second criterion of nonexcludability is the ability of the recipient to consume the good. The cost (especially the political cost involved) of excluding those lacking in capability may be much less than to exclude those with expertise and capability, since they would be much more likely to demand equal access. Thus, nonexcludability may be a function of the capabilities of the consumers, and not necessarily an intrinsic property of the good itself. The consumption criterion of nonexcludability is used in arguments put forward by LDCs, pointing out that while present regime rules do not bar them from access to the orbit/spectrum resource, their lack of technical and financial means does. Thus, their insistence on the provision of "equitable access guaranteed in practice." [5]

The capacity to efficiently exploit a collective good may vary greatly among recipients. If these consumers then posed additional costs in terms of rivalry and congestion through their inefficient exploitation, it may be more economical to exclude them from direct consumption in the long run by having them use the space telecommunications networks of more advanced states or firms. At times, this is heard as an argument from the "space-haves," who point to INTELSAT and other global networks as the preferred means for LDC access.

3. *The Problem of Regime Production*

If a good is indivisible and nonexcludable, *rational, self-interested individuals will not act to achieve their common or group interests.* The difficulty in excluding non-contributors from enjoying the benefits of a collective good leads to that good's underproduction due to the so-called "free-rider problem."

The collective goods perspective is especially applicable to the policy-making environment of the international system [6]. The lack of a higher authority in the international system corresponds to the inability of a producer (the regime) to exclude non-contributors from enjoying the benefits (or "bads") of the good. Therefore, international regimes, as providers of collective goods, typically underproduce. This "explains" the propensity of cooperative-use regimes, such as COPUOS and the ITU, to depend on consensus-building procedures, because the nonexcludability of the regime-produced goods lowers exit costs. Successful regimes, again COPUOS and the ITU, provide excludable benefits to its members in the form of technical information which many countries would never have the opportunity to receive through any other vehicle.

The Orbit/Spectrum Regime

The cooperative-use orbit/spectrum regime, embodied in the ITU, is an institutionalization of rights and rules governing access and use of *res communis* resources for global, regional, and national satellite networks. The regime promotes efficient and interference-free spectrum and orbital slot utilization, which, in turn, allows production of the demanded good — satellite communications channels. The regime is produced to the extent countries perceive they must collaborate in order to be able to gain access to interference-free satellite networks. There is a supply and demand mechanism determining the degree of international collaboration and cooperation. As such, regimes are "demanded" or "supplied."

The *demand* for an international regime results from subjectively based policy-maker perceptions regarding actions taken in pursuit of an idiosyncratically defined national interest. In other words, many countries may demand, or decide to participate in a regime, for many different reasons. The *supply* of an international regime, in contrast, is imposed upon countries from factors outside of their individual control. These factors of regime supply are unambiguous in that countries are responding to the same stimulus [7]. The spectrum is the "supply factor" for the ITU regime.

1. The Resources

The radio spectrum is a very special type of global resource. It is ubiquitous and universally accessible. It is uniquely nonexcludable. Its use or misuse involves far-reaching externalities. Two users on or near the same frequency may cause interference to each other, so as to make communications impossible. The power, or ability, to generate electromagnetic waves and the use or misuse of the spectrum is uniformly and universally distributed. Potentially, anyone can "jam" anyone else. The spectrum-imposed balance of power closely approximates what Keohane calls the "international context."

> Two features of the international context are particularly important: world politics lacks authoritative governmental institutions, and is characterized by pervasive uncertainty. Within this setting, a major function of international regimes is to facilitate the making of mutually beneficial agreements among governments so that the structural condition of anarchy does not lead to a complete "war of all against all." [8]

The powerlessness of any state to prevent another from generating electromagnetic waves represents the spectrum's nonexcludability and is the strongest incentive for regime participation. Unregulated use of the ubiquitous (i.e., indivisible), spectrum resource, brings with it interference externalities indicative of a "market failure." Thus, *the spectrum resource resembles a collective good in the way it must be used for efficient utilization.*

Users of the spectrum resource realize that to communicate efficiently, or if at all, they must cooperate with each other. Cooperation requires agreements, communications, information, and a site for negotiations, all of which add up to enormous "transaction costs" if each state were to attempt to coordinate spectrum use with all other states through bilateral treaties or agreements. ITU Administrative Radio Conferences reduce transaction costs while encouraging cooperation between its over 158 members.

In contrast to the spectrum resource, the geostationary orbit is not universally accessible. Very few countries have the technical and financial means with which to gain direct access to the orbit. Nor is it an indivisible resource. Most crucially, occupation of an orbital slot in a saturated portion of the geostationary orbit pre-empts another from using the same location with the same service area. This is presently the situation for the C- and Ku-Band slots for North America. The interference limitations of the spectrum are exacerbated by the spatial constraints posed by the geographical service areas, satellite spacing, antenna radiation patterns, and the size of ground segment antennas, among other factors. In sum, the geostationary orbit is a spatially fixed area outside the jurisdictional prerogatives of any single state, making its use and management a policy concern of all states.

2. Tragedy of the Orbit/Spectrum Commons

As reviewed in Chapter 3, there is a contradiction between the open and free use regime dictated by the Outer Space Treaty [9] and the constraints on use imposed by properties of the resources. Is "free and open use" a valid legal doctrine when applied to the limited and increasingly crowded orbit/spectrum resource? The conflict between the individual and collective interests, as posed by the collective good attributes of the orbit/spectrum resource, resembles the policy dilemma in Garret Hardin's "Tragedy of the Commons." [10]

In the "Tragedy" metaphor, the interest of the community is subjugated by each individual pursuing his or her own self-interest. This is the "free-rider" phenomenon which causes regime underproduction. The ITU regime provides for efficient orbit/spectrum use, yet a rational state can take advantage of this by establishing another satellite system, without losing anything. However, everyone "pays" in terms of additional crowding and interference in the geostationary orbit, which raises the costs to the next user, but the free-rider's costs are still lower than the benefit realized through the satellite system. As Kenneth Arrow pointed out,

> It is certainly a matter of common observation, perhaps most especially in the field of international relations, that mutually advantageous agreements are not arrived at because each party is seeking to engross as much as possible of the common gain for itself. [11]

The nonexcludability or free access characteristic of the spectrum side of the orbit/spectrum resource compels cooperation; yet the regime is underproduced as is evidenced by the increasing politicization and polarization of negotiations. This indicates that although the resource-imposed supply incentives remain unchanged, a factor on the demand side has shifted, making free-riding more feasible. In other words, constraints and opportunities for access to satellite communications channels and networks are changing. This brings us to a discussion of the actual satellite channels and their distribution through INTELSAT.

Satellite Channels: Dimensions of a Collective Good

Martin Rothblatt defines the value of satellite communications and its use of the orbit/spectrum resource as the utility value in providing a channel or pathway for the relaying of information between communicators. [12]

A communications pathway is measured in three dimensions: depth, distribution, and directionality. *Depth* is defined and measured as "the message volume [that a telecommunications channel] can conduct per unit time." The *distribution* of a satellite communications channel is the "mean distance between communicators." The dimension of *directionality* indicates whether information is being sent in primarily one direction as with Direct Broadcast Satellites or two-way message exchanges as in telephone service between individuals or teleconferencing between groups of people. [13]

The three dimensions of the communications channel allow us to determine its "mix" between indivisibility and nonexcludability.

1. Depth and Indivisibility

Indivisibility is represented by the dimension of channel depth where increased message carrying capacity and connectivity offered by communications satellite technology and networks are (1) distance insensitive, and (2) are potentially available to any communicators on earth. The cost of satellite communications is distance-insensitive. This corresponds to the property of indivisibility or jointness of supply, where there is no additional cost in extending satellite service to the next or last user within the satellite's service area. For these reasons satellite communications is potentially available to all communicators on earth, with "connectivity" or jointness cost curves approaching zero.

1.1 Distribution, Directionality, and Free Access

Channel distribution and directionality are properties most closely associated with free access (or nonexcludability), characteristic of a collective good. The INTELSAT network was based upon early satellite technology that was feasible only through international cooperation. In its early stages of technical and administrative development, the global INTELSAT network was a very "pure" collective good (i.e., globally distributed and non-excludable), where access was assured through the indivisibility and nonexcludability of the technology. One global or international system assured open access and a wide distribution of benefits to all participating countries.

Each orbital slot used by an INTELSAT satellite serviced and benefited all countries in that hemisphere. However, the rapid expansion of world utilization of satellite communications for vital national interests encouraged technological innovation that made it possible for countries to achieve the benefits of satellite communications by establishing their own systems that reduced at the same time their dependency costs incurred as participants in INTELSAT. Consequently, we see a proliferation of both regional and national satellite systems exemplified by ARABSAT, EUTELSAT, and other INTELSAT competitors.

A national or regional satellite system has the appearance of a private good for two reasons: (1) its occupation of an orbital slot and use of frequencies in an increasingly crowded orbit/spectrum resource pre-empt their utilization by others, and (2) the benefits from such use are realized only by the operating country; they are "excludable." The use of an orbital slot by a national or dedicated satellite system only benefits the operating country. Technology has changed the distributional properties of satellite networks and, thereby, the incentives for countries to cooperate in the INTELSAT regime, with far-reaching implications for the ITU regime as well.

Synthesis

What does the collective goods perspective tell us? Technology is changing the incentives for forming common-user networks in favor of dedicated national or regional systems. These represent an excludable good of increasing power value for the few countries with the means for direct access to the orbit resource. Although, as Rothblatt points out, the international regime (ITU/INTELSAT) is based on the "maximum dispersion" principle, [14] technology creates exclusionary mechanisms and incentives to limit that dispersion.

In essence, technology has created a free-rider alternative. Excludable regional, national, and dedicated (for a particular class or type of telecommunications service or subscriber) satellite communications networks allow the satellite operator to receive the benefits without the dependency costs incurred as a participant in a global network. INTELSAT is becoming the satellite network for countries that cannot afford direct access, as wealthier countries with the technological and financial means establish their own national and regional systems. [15]

Regional, national, and dedicated satellite systems, as seen from the member state's point of view, reduce dependency costs while at the same time they dissolve free-rider incentives that might operate against even smaller systems. This is seen in the observation that dependency and transaction costs in an organization of 10 members is much less than in INTELSAT with over 100 members. In the smaller system no country thinks it can be a free-rider because of the perception common in smaller groups that each member's contribution is crucial to the production of the good [16]. At the same time, the regional system allows each of its members to be a free-rider against INTELSAT. Each country can "skim off" the bulk of its telecommunications traffic onto the regional satellite, while incurring lower dependency costs commensurate with its amount of global traffic carried by INTELSAT.

The trick to encouraginng contributions to a collective good is to make the good appear as private as possible. Increasingly excludable regional and national satellite systems can be regulated through simultaneous management of the orbit/spectrum resource *on a regional basis.* [17] Incentives for cooperat-on would grow as dependency costs were reduced in the smaller organization. Each region would be free to develop resource planning to fit the satellite systems deemed most useful for achieving each region's telecommunications objectives. Increasingly excludable satellite networks, made possible by technologies with fewer externalities (interference), also allow regional management of the resources. In other words, breaking the ITU into regional organizations for some types of telecommunications may promote cooperation.

Remarks

The impasse over planning, allocations, allotments, and management of the orbit/spectrum resource is a conflict between the interest of the individual *versus* that of the collectivity, a situation resembling Hardin's "Tragedy" metaphor. As Olson states,

> Taken individually, the states of the world are more often than not rational; taken together, they constitute an international system that is usually irrational. They conform, in other words, to central insight of the theory of collective goods: with these goods, unlike others, rational individual behavior normally does not spontaneously lead to a rational collective outcome. Only arrangements designed to give individual states an incentive to act in their common interest can bring a collectively sane result. [18]

The collective good aspects of the spectrum resource compel cooperation between users based upon a distribution of power established by the resource's free accessibility.

In Keohane's framework, the ITU is a "supplied" regime. Any state or individual can produce interference. Ironically, advanced information societies with their own satellite networks are more vulnerable to spectrum coercion, or the jamming of communications satellites. The free and universal accessibility of the spectrum grants all states an effective method of retaliation through intentional jamming. Reciprocity is an effective deterrent, if both sides are mutually and symmetrically vulnerable. It can be shown, therefore, to be in the interest of the space-haves to ensure space have-nots to satellite networks and the orbit/spectrum resource. In this way, all countries are mutually vulnerable to spectrum coercion. The overriding fact that the spectrum must be used as a collective good, is, therefore, a strong incentive for collaboration and cooperation between users.

However, the orbit/spectrum resource regime is politicized for reasons removed from the issue of access. Beyond the question of the distribution of access is the issue of use, and how the benefits should be distributed. INTELSAT is a "demanded" regime, through which participating countries gain direct access to satellite communications channels, and indirect access to the orbit/spectrum resource by utilizing INTELSAT satellites. INTELSAT uses the orbit/spectrum resource and provides a global network of satellite channels to all countries. This network, as we have seen, is itself a collective good. It is also prone to underproduction due to the free-rider phenomenon, which in this case takes the form of competing regional, national, and dedicated satellite systems, exemplified by the recent applicants for the North Atlantic circuits.

A disintegration of INTELSAT into a plethora or proliferation of smaller, less efficient satellite systems would have far-reaching implications and ramifications, not just for the countries dependent on INTELSAT, but also for all orbit/spectrum users. The collective goods model is a useful guide for evaluating possible reforms in the existing regime.

REFERENCES

1. The theoretical assumptions and approaches are examined in depth by the following works: The discussion on supply and demand of regimes is derived from: Bruno S. Frey, "The Public Choice View of International Political Economy," *International Organization* 38:1(Winter 1984):199-224; Robert Keohane, "The Demand for International Regimes," *International Organization* 36(Spring 1982):325-355; Mancur Olson and Richard Zeckhauser, "An Economic Theory of Alliances," *The Review of Economics and Statistics*, 48(1966):266-279; John Ruggie, "Collective Goods and Future International Collaboration," *American Political Science Review*, 66(1972):874-893; Bruce Russett

and John Sullivan, "Collective Goods and International Organization," *International Organization*, 25(1971):845-865; Todd Sandler and Jon Cauley,"The Design of Supranational Structures: An Economic Perspective," and Todd Sandler and William Schulze, "Explorations in the Economics of Outer Space," in Sandler (ed.), *The Theory and Structures of International Political Economy*, Boulder: Westview Press 1980; Alex G. Vicas, "Efficiency, Equity and the Optimum Utilization of Outer Space as a Common Resource," in *Annals of Air and Space Law* 5(1980):589-609; Per M. Wijkman and Clas G. Wihlborg, "Global Use and Regulation of Space Activities under the Common Heritage Principle," paper presented at International Conference on Doing Business in Space: Legal Issues and Practical Problems, November 12-14, 1981, Washington, D.C.; and, "Outer Space Resources in Efficient and Equitable Use: New Frontiers for Old Principles," in *The Journal of Law and Economics* 24(April 1981):23-44.

2. See Edward L. Morse, "Managing International Commons," in *Journal of International Affairs*, 31(1977):4-5.
3. Richard Kimber, "Collective Action and the Fallacy of the Liberal Fallacy," in *World Politics*, 23(January 1981):179.
4. Ruggie, p. 887.
5. International Telecommunication Convention, Article 33, 1973, and Resolution BP, Final Acts, WARC-79.
6. Duncan Snidal, "Public Goods, Property Rights, and Political Organization," in *International Studies Quarterly*, 24(December 1979):532-566.
7. Keohane, p. 332.
8. *Ibid.*
9. Treaty on Principles Governing the Activities of States in the Exploration and Use of Outer Space, Including the Moon and Other Celestial Bodies, January 27, 1967, 18 U.S.T. 2410, T.I.A.S. 6347, 610 U.N.T.S. 205.
10. Garret Hardin, "The Tragedy of the Commons", in *Science*, 162(13 December 1968):1243-48.
11. Kenneth Arrow, "The Organization of Economic Activity: Issues Pertinent to the Choice of Market versus Non-market Allocation", quoted in Alex G. Vicas, "Efficiency, Equity and the Optimum Utilization of Outer Space as a Common Resource," in *Annals of Air and Space Law*, 5(1980):589-609, fn. 14.
12. M. Rothblatt, "Satellite Communication and Spectrum Allocation" in *American Journal of International Law* 76(1)January 1982:56-77; "A Jurimetric Framework for the International Allocation and Economic Development of the Orbit/Spectrum Resource;" *Proc. of the 24th Colloquium on the Law of Outer Space*, 1981:79-86; "The Impact

of International Satellite Communications Law Upon Access to the Geostationary Orbit and the Electromagnetic Spectrum" in *Texas International Law Journal* 16(1)1981:207-244.

13. Rothblatt, "Impact," 1981:209.
14. Rothblatt, Proceedings, 1981:80.
15. As Frey writes,

    [L]arge international organizations have become less effective because the share of the benefits taken by the dominant country (formerly the United Kingdom, now the United States) has decreased, leading to less cooperation. Instead of a leader providing international public goods partly in its own interest, the dominant force is now a group of relatively small countries, each of which is unwilling to provide public goods. (Frey, p. 217.)

16. M. Olson, *The Logic of Collective Action*, (1965).
17. This was one of the conclusions reached by a study done by the US Congress, Office of Technology Assessment, *Radiofrequency Use and Management: Impact from WARC-79*, 1982, p. 118.
18. M. Olson, "Increasing the Incentives for International Cooperation," *International Organization* 25(1971):873.

GLOSSARY

ABBREVIATIONS AND TERMINOLOGY

CCIR
: International Radio Consultative Committee

CCITT
: International Telephone and Telegraph Consultative Committee

COPUOS
: Committee on the Peaceful Uses of Outer Space

EMP
: Electromagnetic Pulse

FCC
: Federal Communications Commission

FDMA
: Frequency-Division Multiple Access

Geoplat
: Geostationary Platform

Group 77
: Voting Bloc of Less Developed Countries

GSO
: Geostationary Orbit

HDCs
: Highly Developed Countries

IFRB
: International Frequency Registration Board

INMARSAT
: International Maritime Satellite Organization

INTELSAT
: International Telecommunications Satellite Organization

INTERSPUTNIK
: Soviet Bloc Global Satellite System

ISDN
: Integrated Services Digital Network

ITU
International Telecommunication Union

LDCs
Less Developed Countries

MNCs
Multinational Corporations

"North"
Industrialized Countries (HDCs)

PTT
Post, Telegraph, and Telephone Administration or Ministry

RARC
Regional Administrative Radio Conference

"South"
Developing Countries (LDCs)

TBDF
Transborder Data Flows

TDMA
Time-Division Multiple Access

WARC
World Administrative Radio Conference

BIBLIOGRAPHY

BOOKS and ARTICLES IN BOOKS

Bloomfield, Lincoln P., ed., *Outer Space: Prospects for Man and Society*. New York: Praeger, 1968.

Brown, Seyom, *et al.*, *Regimes for the Ocean, Outer Space, and Weather*. Washington: Brookings Institute, 1977.

Christol, Carl Q., *The Modern International Law of Outer Space*. New York: Pergamon Press, 1982.

Codding, George A., *The International Telecommunication Union: An Experiment in International Cooperation*. New York: Arno Press, 1972.

Codding, George A., and A.M. Rutkowski, *The ITU in a Changing World*. Dedham: Artech House, 1982.

Curtin, D.J., ed., *Trends in Communications Satellites*. New York: Pergamon Press, 1979.

Deutsch, Karl, *The Nerves of Government*. New York: The Free Press, 1963.

Dizard, Wilson, *The Coming Information Age: An Overview of Technology, Economics, and Politics*. New York: Longman, Inc., 1982.

Dolman, Antony J., ed., *Global Planning and Resource Management: Toward International Decision-Making in a Divided World*. New York: Pergamon Press, 1980.

Dordick, Herbert S., ed., *Proceedings of the Sixth Annual Telecommunications Policy Research Conference*. Lexington: D.C. Heath and Company, 1979.

Engelhard, Helmut, *Satellitendirektfersehen, neue Technologie fuer einen besseren internationalen Informationsfluss?: Die voelkerrechtliche Kontroverse zwischen Informationsfreiheit und Staatsouveraenitaet*. Frankfurt: Verlag Peter Lang GmbH, 1978.

Feigenbaum, Edward A. and Pamela McCorduck, *The Fifth Generation*. New York: Addison-Wesley, 1983.

Galloway, Jonathan F., *The Politics and Technology of Satellite Communications*. Lexington: D.C. Heath and Company, 1972.

Gerbner, George, and Marsha Siefert, editors, *World Communications: A Handbook*. New York: Longman, 1984.

Goldsen, Joseph M., *Outer Space in World Politics*. New York: Praeger, 1963.

Gould, R.G., ed., *Communication Satellite Systems: An Overview of the Technology*. IEEE Press, 1976.

Gunther, Jonathan F., *The U.S. and the Debate on the World Information Order*. Washington, D.C.: Academy for Educational Development, Inc., 1978.

Haley, Andrew G., *Space Law and Government*. New York: Appleton-Century-Crofts, 1963.

International Telecommunication Union. *From Semaphore to Satellite*. Geneva: International Telecommunication Union, 1965.

Jaffe, Leonard, ed., *Satellite Communications in the Next Decade: Proceedings of the 14th Goddard Memorial Symposium*. San Diego: Univelt, 1977.

Jansky, Donald M., *World Atlas of Satellites*. Dedham: Artech House, 1983.

Jansky, Donald M. and Michel C. Jeruchim, *Communication Satellites in the Geostationary Orbit*. Dedham: Artech House, 1983.

Jasani, Bhupendra, ed., *Outer Space - A New Dimension of the Arms Race*. Stockholm: Stockholm International Peace Research Institute, 1982.

Jasentuliyana, Nandasiri, and Roy S.K. Lee, *Manual on Space Law: Volumes I and II*. Dobbs Ferry: Oceana Publications, Inc., Alphen aan den Rijn: Sijthoff and Noordhoff, 1979.

Kash, Don E., *The Politics of Space Cooperation*. West Lafayette: Purdue University Studies, 1967.

Kildow, Judith T., *INTELSAT: Policy-Maker's Dilemma*. Lexington: D.C. Heath, 1973.

Kinsley, Michael E., *Outer Space and Inner Sanctums: Government, Business, and Satellite Communication*. New York: John Wiley and Sons, 1976.

Lay, Houston S. and H.J. Taubenfeld, *The Law Relating to the Activities of Man in Outer Space*. Chicago: University of Chicago Press, 1970.

Leive, David, *International Telecommunications and International Law: The Regulation of the Radio Spectrum*. Dobbs Ferry: Oceana Publications, 1970.

Levin, Harvey, *The Invisible Resource: Use and Regulation of the Radio Spectrum*. Baltimore: The Johns Hopkins Press, 1971.

McWhinney, Edward, ed., *The International Law of Communications*. Dobbs Ferry: Oceana Publications, 1971.

Magnant, Robert S., *Domestic Satellite: An FCC Giant Step*. Boulder: Westview Press, 1977.

Martin, James, *Communications Satellite Systems*. Englewood Cliffs: Prentice Hall, 1978.

Musolf, Lloyd D., ed., *Communications Satellites in Political Orbit*. San Francisco: Chandler, 1968.

Nordenstreng, Kaarle and Herbert Schiller, *National Sovereignty and International Communications*. Norwood: Ablex, 1979.

Oettinger, Anthony G., Paul J. Berman, and William H. Read, *High and Low Politics: Information Resources for the 1980s*. Cambridge: Ballinger, 1981.

Pelton, Joseph N., *Global Communications Satellite Policy: INTELSAT, Politics, and Functionalism*. Mt. Airy: Lomond Books, 1974.

Pelton, Joseph N., and M.S. Snow, eds., *Economic and Policy Problems in Satellite Communications*. New York: Praeger, 1977.

Porat, M., "Communications Policy in an Information Society," in Glenn Robinson, ed., *Communications for Tomorrow: Policy Perspective for the 1980s*. New York: Praeger, 1978.

Queeney, Kathryn M., *Direct Broadcast Satellites and the United Nations*. Alphen ann den Rijn: Sijthoff and Noordhoff International Publishers B.V., 1978.

Report of the Twentieth Century Fund Task Force on International Satellite Communications, *Communicating By Satellite*. New York: Twentieth Century Fund, Inc., 1969.

Sanders, Ralph, *International Dynamics of Technology*. Westport: Greenwood Press, 1983.

Sandler, Todd, and Jon Cauley, "The Design of Supranational Structures: An Economic Perspective," and Todd Sandler and William Schulze, "Explorations in the Economics of Outer Space," in Sandler, ed., *The Theory and Structures of International Political Economy*. Boulder: Westview Press, 1980.

Sandler, Todd, and William Schulze, "Explorations in the Economics of Outer Space," in Sandler, ed., *The Theory and Structures of International Political Economy*. Boulder: Westview Press, 1980.

Schauer, William H., *The Politics of Space: A Comparison of the Soviet and American Space Programs*. New York: Holmes and Meier, 1976.

Schiller, Herbert I., *Who Knows: Information in the Age of the Fortune 500*. Norwood: Ablex, 1981.

Signitzer, Benno, *Regulation of Direct Broadcasting Satellites: The U.N. Involvement*. New York: Praeger, 1976.

Singh, Indu B., ed., *Telecommunications in the Year 2000: National and International Perspectives*. Norwood: Ablex, 1983.

Smith, Anthony, *The Geopolitics of Information: How Western Culture Dominates the World*. New York: Oxford University Press, 1980.

Smith, Delbert D., *International Telecommunication Control: International Law and the Ordering of Satellite and Other Forms of International Broadcasting.* Leyden: Sijthoff, 1969.

Smith, Delbert D., *Communication Via Satellite: A Vision in Retrospect.* Boston: Sijthoff, 1976.

Smith, Delbert D., *Space Stations: International Law and Policy.* Boulder: Westview Press, 1979.

Smythe, Dallas W., *Dependency Road: Communications, Capitalism, Consciousness and Canada.* Norwood: Ablex, 1981.

Snow, Marcellus S., *International Commercial Satellite Communications: Economic and Political Issues of the First Decade of INTELSAT.* New York: Praeger, 1976.

Stockholm International Peace Research Institute, *Communication Satellites.* Stockholm: Almqvist and Wiksell, 1969.

Stockholm International Peace Research Institute, *Outer Space —Battleground of the Future?* Stockholm: Stockholm International Peace Research Institute, 1978.

White, Irvin L., *Decision-Making for Space: Law and Politics in Air, Sea, and Outer Space.* West Lafayette: Purdue University Studies, 1970.

Wicklein, John, *Electronic Nightmare: The New Communications and Freedom.* New York: Viking Press, 1981.

DOCUMENTS

1. TREATIES

1963: *Declaration of Legal Principles Governing the Activities of States in the Exploration and Use of Outer Space.* United Nations General Assembly Resolution 1962 (XVIII), December 13, 1963.

1967: *Treaty on the Principles Governing the Activities of States in the Exploration and Use of Outer Space, Including the Moon and Other Celestial Bodies* (*Outer Space Treaty*). Open for signature January 27, 1967, 18 U.S.T. 2410. T.I.A.S. 6347. 610 U.N.T.S. 205, entered into force for the U.S. October 10, 1967.

1968: *Agreement on the Rescue of Astronauts, the Return of Astronauts and the Return of Objects Launched into Outer Space.* Open for signature April 22, 1968, 19 U.S.T. 7570, T.I.A.S. 6599, 672 U.N.T.S. 119, entered into force for the U.S. December 3, 1968.

1972: Convention on International Liability for Damage Caused by Space Objects. Open for signature March 29, 1972, 24 U.S.T. 2389, T.I.A.S. 7762, entered into force for the U.S. October 9, 1973.

1975: *Registration of Objects Launched Into Outer Space.* Done, January 14, 1975, 28 U.S.T. 695, T.I.A.S. 8480, entered into force for the U.S. September 15, 1976.

1979: *Agreement Governing the Activities of States on the Moon and Other Celestial Bodies* (*Moon Treaty*), has not entered into force. UN Document A/AC.105-L113/Add 4.

INTELSAT Agreement, signed August 20, 1971. T.I.A.S. 7532.

International Telecommunication Union, *Final Acts*, World Administrative Radio Conference, 1979.

International Telecommunication Union, *Final Acts.* World Administrative Radio Conference — Space Telecommunications, 1971.

International Telecommunication Union, *International Telecommunication Convention.* Malaga-Torremolinos, 1973.

International Telecommunication Union, *International Telecommunication Convention.* Nairobi, 1982.

2. ITU DOCUMENTS

International Telecommunication Union, *World Communications: New Horizons/New Power/New Hope.* Geneva: Telecom '83, 1983.

International Telecommunication Union, "Contribution of the International Radio Consultative Committee to Unispace '82." Background paper A/CONF.101/BP/IGO/9. April 1982.

International Telecommunication Union, Administrative Council Report to the Plenipotentiary Conference. "Review of the State of Telecommunications Services in the Least Developed Countries and Concrete Measures for Telecommunications Development." Doc. No. 48-E. Geneva, May 1982.

International Telecommunication Union, CCIR. *Proposed CCIR Conference Preparatory Meeting (Special Joint Study Group Meeting) for Regional Administrative Radio Conference for Planning the Broadcasting Satellite Service in Region 2.* Administrative Circular A.C./241. Add. 2. December 18, 1981.

International Telecommunication Union. CCIR, "Provisional Technical Report for WARC 84." Doc. 4/286-E. Report from IWP 4/1.

International Telecommunication Union. "The Missing Link: Report of the Independent Commission for World Wide Telecommunications Development." (Maitland Commission Report) December 1984, Geneva.

International Telecommunication Union. *Twenty-first Report by the International Telecommunication Union on Telecommunication and the Peaceful Uses of Outer Space.* Booklet No. 30. Geneva, 1982.

International Telecommunication Union. *Twenty-second Report by the International Telecommunication Union on Telecommunication and the Peaceful Uses of Outer Space.* Booklet No. 31. Geneva: 1983.

3. OTHER DOCUMENTS

Department of Defense, Directive: Military Satellite Communications Systems Organization MILSATCOM. No. 5105.44, October 9, 1973.

Federal Communications Commission, "ITU-WARC/ORB-85 Second Advisory Committee Report." January 31, 1985.

Organization of Economic Cooperation and Development. *Telecommunications: Pressures and Policies for Change.* Paris: OECD, 1983.

Pierce, W. and N. Jequier, "Telecommunications and Development: General Synthesis Report on the Contribution of Telecommunications to Economic and Social Development," ITU/OECD, Geneva and Paris, 1982.

Public Service Satellite Consortium. "Satellite Communications for the Pacific Islands." December 1982, Washington, D.C.

United Nations, Unispace '82, "Background Paper: Relevance of Space Activities to Economic and Social Development." A/CONF.101/BP/8. April 7, 1981.

Unispace '82, "Declaration of the Group of 77," A/CONF.101/5. August 13, 1982.

United States Congress. Office of Technology Assessment. *Radiofrequency Use and Management: Impacts from the World Administrative Radio Conference of 1979.* January 1982.

United States Congress. Office of Technology Assessment. *Civilian Space Policy and Applications* Washington, D.C.: Government Printing Office, 1982.

United States Congress. Office of Technology Assessment. *Unispace '82: A Context for International Cooperation and Competition.* Washington, D.C.: Government Printing Office, 1983.

ARTICLES

1. ARTICLES ON SPACE POLITICS

Al-Mashat, Ali, "The Arab Satellite Communications System," *Proceedings of the AIAA 9th Satellite Communications Conference.* San Diego, California, March 1982, pp. 187-191.

Arab Satellite Communications Organization (ARABSAT). *Background Paper: Arab Satellite Communications Organization.* A/CONF.101/BP/I-GO/4, submitted 8 July 1981, distributed at Unispace '82, August 1982, Vienna.

Arnopoulos, Paris, "The International Politics of the Orbit / Spectrum Issue." *Annals of Air and Space Law* 7(1982):215-239.

Bailey, Robert, "Increase in Capacity, Sophistication Aids New Regional Satellite Systems," *International Herald Tribune*, September 23, 1982, p. 8.

Branscomb, Anne W., "Global Infrastructure for Information Transport." Paper presented at Forum '83, October 29, 1982, Geneva.

"Congress May Move Radio Marti Closer to Reality." in *Broadcasting*, June 13, 1983, p. 42.

Brooks, Harvey, "Managing the Enterprise in Space." *Technology Review*. April 1983, pp. 39-47.

"Telecommunications: The Global Battle." *Business Week*. October 24, 1983, pp. 62-65.

Brunt, Peter and Alan I. Naylor, "Telecommunications and Space." *Futures* 14:1(1982):417-434.

Butler, Richard E., "Satellite Communications for Developing Countries." Monograph presented at the United Nations Interregional Seminar on Space Applications in Preparation for Unispace '82, Addis Ababa, 18 June 1982.

Butler, Richard E., "Application of Space Telecommunications for Development — Service Prospects for the Rural Areas." Paper presented at Unispace '82, Vienna, August 1982.

Caming, H.W. William, "The Protection of Transborder Data Information in the United States." Paper presented at Forum '83, October 29, 1983, Geneva.

Chamoux, Jean Pierre, "Borderlines to Free Flow: Perspectives of the International Business." Paper presented at Forum '83, October 29, 1983, Geneva.

Cook, James, "The Molting of America," in *Forbes*, November 22, 1982, p. 163.

Cowlan, Bert, "Internationally Organizing for Space: Historical Perspectives." Paper presented to the American Bar Association,*et al.* Conference on "Doing Business in Space," Washington, D.C., November 18-21, 1981.

Cowlan, Bert, "Unispace '82: A Retrospective." Paper submitted November 9, 1982, to Office of Technology Assessment, and comments during Unispace Technical Memorandum Committee Meeting, December 2, 1982, Washington, D.C.

Cruise O'Brien, Rita, and G.K. Helleiner, "The Political Economy of Information in a Changing International Economic Order." *International Organization* 34:4 (Autumn 1980).

Denis-Lempereur, Jacqueline, "La vraie reforme nous vient du ciel." *Science et Vie*. May 1982, pp. 72-201.

"Fernsehan: Gold im All," *Der Spiegel*, (March 2, 1981), pp. 109-110. Lexington: D.C. Heath and Company, 1979.

Dizard, Wilson P., "Space WARC and the Role of Internatinal Satellite Networks." Paper presented at the Center for Strategic and International Studies, Georgetown University, Washington, D.C. August 1984.

DuCharme, E.O., R.R. Bowen, and M.J.R. Irwin, "The Genesis of the 1985-87 ITU World Administrative Radio Conference on the Use of the Geostationary Satellite Orbit and the Planning of Space Services Utilizing It." *Annals of Air and Space Law* 8(1982):261-281.

Duigou, Michel, "Le programme Arabsat." *Aerospatiale* No. 123, October 1982, p. 23.

Edelson, Burt, Richard Marsten, and Walter Morgan, "Greater Message Capacity for Satellites." *IEEE Spectrum*. March 1982, pp. 56-64. New York: Praeger, 1963.

Goeschel, Wilhelm, "Direct Television via Satellite." *Endeavour*. No. 4, (1981), pp. 160-165.

Grabhorn, Edgar A., Director of World Telecommunications Information Program, Arthur D. Little, Inc., quoted in *Inteltrade*, October 15-30, 1982, p. 1.

Hagelin, Theodore M., "Prior Consent or The Free Flow of Information over International Satellite Radio and Television: A Comparison and Critique of U.S. Domestic and International Broadcast Policy." *Syracuse Journal of International Law and Commerce* 8:2(1981):265-320.

Halloran, Richard, "United States Turning Its Attention to the New Theater of Military Operations: Space," in *International Herald Tribune*, October 20, 1982.

Hudson, Heather E., Andrew P. Hardy, and Edwin B. Parker, "Impact of Telephone and Satellite Earth Stations Installations on GDP," *Telecommunications Policy* 6:4 December 1982, pp. 300-307.

Karas, Tom, comments broadcast by the Public Broadcasting System in a program entitled, "The Race for the High Ground," broadcast on "Frontline," April 13, 1983.

Keohane, Robert, "The Demand for International Regimes," in *International Organization* 36(Spring 1982):325-355.

Lenorovitz, Jeffrey M., "Luxembourg Set to Authorize Company to Operate Coronet TV Satellite." *Aviation Week and Space Technology*. January 28, 1985, pp. 92-93.

Locksley, Gareth, "The Political Economy of Satellite Business." *Telecommunications Policy* 7(September 1983):195-203.

Logsdon, Tom, "Orbiting Switchboards." *Technology Illustrated.* October-November 1981, pp. 55-62.

Logsdon, Tom, "Unsnarling Signals in Space." *Technology Review*, August/September 1982, pp. 14.

McCaskey, Scott D, "Satellite Communications — Stepping Stones for Developing Nations." *Satellite Communications*, July 1980, pp. 22-25.

Madec, Alain, "The Political Economy of Information Flows." *Intermedia* 9(March 1981):29-32.

Marchand, Montigny, "The Impact of Information Technology on International Relations." *Intermedia* 9(November 1981):12-15.

Martinez, Larry F., "The Orbit/Spectrum Resources and Regime as Collective Goods: Perspective for Cooperation." *26th Proceedings of the Colloquium on Space Law*, International Institute of Space Law, Budapest, October 1983.

Martinez, Larry F., "Butler Ushers the ITU in the 1980s," *Satellite Communications*, March 1983, pp. 42-45.

May, Michael, "War or Peace in Space." Monograph published by the California Seminar on International Security and Foreign Policy No. 93, March 1981, p. 2.

Miles, Edward, "Transnationalism in Space: Inner and Outer," *International Organization.* 25(1971):604.

Miles, Ian, and Michiel Schwarz, "Alternative Space Futures: The Next Quarter-Century." *Futures* 14:1(1982):461-483.

Morgan, Walter L., "Satellite Economics in the 1980s," *Satellite Communications*, (January 1980), pp. 26-29.

Morse, Edward L., "Managing International Commons." *Journal of International Affairs* 3:1(1977):1-21.

Naeslund, Ruben, "Some Regulatory Aspects on Matters which Will Be Treated by Forthcoming ITU Conferences." Paper presented at Forum '83, October 29, 1983, Geneva.

National Aeronautics and Space Administration. *Proceedings of the Conference of the Law of Space and of Satellite Communications.* Washington, D.C.: NASA Scientific and Technical Information Division, Government Printing Office, 1964.

O'Neill, John., "Technical Assistance to Developing Countries." *Telecommunications*, April 1981, pp. 67-79.

Olson, Mancur, and Richard Zeckhauser, "An Economic Theory of Alliances." *The Review of Economics and Statistics.* 48(1966):266-279.

Pirard, Theo, "Intersputnik: The Eastern "Brother" of Intelsat," *Satellite Communications*, August 1982, pp. 38-44.

Ploman, Edward, "National Needs in an International Communications Setting." Paper presented at Pacific Telecommunications Conference, January 16, 1983, Honolulu.

Ploman, Edward, "The Whys and Wherefores of International Organizations." *Intermedia* 8(July 1980):6-11.

Pritchard, Wilbur L., and Charles A. Kase. "Getting Set for Direct-Broadcast Satellites." *IEEE Spectrum*. August 1981, pp. 22-28.

Ruggie, John, "Collective Goods and Future International Collaboration." *American Political Science Review* 66(1972):874-893.

Russett, Bruce, and John Sullivan, "Collective Goods and International Organization." *International Organization* 25(1971):845-865.

Russotto, Jean, "Data Protection in Europe: The Council of Europe Convention, a Legal Practitioner's View." Paper presented at Forum '83, October 29, 1983, Geneva.

Sauvant, Karl P., "Transborder Data Flows and the Developing Countries." *International Organization* 37(Spring 1983):359-371.

Schultz, James B., "Reliable Survivable Satellites Seen as Key Link in United States Security," in *Satellite Communications* June 1980, p. 26.

Shackman, A.D., "Are Satellites in the Stars for Europe?" *Data Communications*, April 1983, pp. 158-163. Norwood: Ablex, 1981.

Snidal, Duncan, "Public Goods, Property Rights, and Political Organization." *International Studies Quarterly* 24 (December 1979):532-566.

Tong, David, "An Evaluation of Large Space Platforms." *Canadian Aeronautics and Space Journal* 26:4(Fourth Quarter, 1980):279-280.

"Earth Stations for Rural Communications Systems." *Telephony*, July 25, 1983, pp. 7-9.

Vartabedian, Ralph, "Satellites Spur a World Revolution: Launch 20 Years Ago Changed Course of Communications." *Los Angeles Times*, July 26, 1983, part IV, pp. 1-4.

Vicas, Alex G., "Efficiency, Equity and the Optimum Utilization of Outer Space as a Common Resource." *Annals of Air and Space Law* 5(1980):589-609.

Voute, Caesar, "Space for Whom?" *Futures* 14:1(1982):448-460.

White, Wade, and Morris Holmes, "The Future of Commercial Satellite Telecommunicatons." *Quest* 2:1(Spring 1978):46-69.

Wijkman, Per M., and Clas G. Wihlborg, "Global Use and Regulation of Space Activities under the Common Heritage Principle." Paper presented at International Conference on Doing Business in Space: Legal Issues and Practical Problems, November 12-14, 1981, Washington, D.C.

Wijkman, Per M., and Clas G. Wihlborg, "Outer Space Resources in Efficient and Equitable Use: New Frontiers for Old Principles," in *The Journal of Law of Economics* 24(April 1981):23-44.

"Toward the Wired Society," *World Business Weekly*, June 8, 1981, p. 31.

"Elektroschocks aus dem All," *Die Zeit*, January 15, 1982.

2. ARTICLES ON SPACE LAW

Arzinger, R., "Use of the Geostationary Orbit: The Freedom of Outer Space and the Geostationary Orbit." *21st Proceedings of the Colloquium on Space Law*, International Institute of Space Law, (1979):12-16.

Boeckstiegel, Karl-Heinz, "Settlement of Disputes on International Regimes Applicable to Space Activitites," *23rd Proceedings of the Colloquium on Space Law*, International Institute of Space Law, (1980):123-127.

Bourely, M.G., "The Legal Status of the European Space Agency," *23rd Proceedings of the Colloquium on Space Law*, International Institute of Space Law, 1980, unnumbered.

Bueckling, Adrian, "Rechtsprobleme des Synchronkorridors." *Zeitschrift fur Luft- und Weltraumsrecht Fragen* 76:27(1978):76-85.

Bueckling, Adrian, "Strategy of Semantics and the 'Mankind Provisions' of the Space Treaty," *Journal of Space Law* 7:1(1979):15-22.

Busak, Jan, "Space Telecommunications at Present and in Future," *22nd Proceedings of the Colloquium on Space Law*, International Institute of Space Law, (1979):29-311

Busak, Jan, "The Geostationary Orbit-International Cooperation or National Sovereignty?" *Telecommunication Journal* 45(1978):167-170.

Cassidy, Daniel E., "International Space Cooperative: Participation of Private Enterprises." *23rd Proceedings of the Colloquium on Space Law*, International Institute of Space Law, (1980):133-137.

Chasia, Henry, "Intelsat's Utilization Orbit, Spectrum and Technology to Meet System Requirements in the 1990s." *AIAA 9th communications Satellite Conference*, San Diego, March 7-11, 1982, pp. 192-201.

Christol, Carl Q., "Inventory of Space Activities: Legal Aspects." Unpublished paper, University of Southern California, 1981.

Christol, Carl Q., "The 1974 Brussels Convention Relating to the Distribution of Program-Carrying Signals Transmitted by Satellites: An Aspect of Human Rights." *Journal of Space Law* 6(1978):19-35.

Christol, Carl Q., "The Geostationary Orbital Position as a Natural Resource of the Space Environment." *Netherlands International Law Review* 26:1(1979).

Christol, Carl Q., "National Claims for Using/Sharing of the Orbit/Spectrum Resource." *25th Proceedings of the Colloquium on Space Law*, International Institute of Space Law, (1982).

Christol, Carl Q., "Telecommunications, Outer Space, and the New International Information Order." *Syracuse Journal of International Law and Commerce* 8:2(1981)):343-364.

Cocca, Aldo, Armando, "The Geostationary Orbit, Focal Point of Space Telecommunication Law." *Telecommunications Journal* 45(1978):171-73.

Colino, Richard, "International Cooperation Between Communications Sa-
Doyle, Stephen E., "INMARSAT: The International Maritime Satellite Organization — Origins and Structure." *Journal of Space Law 5:1 (1977):*65-92.

Courteix, Simone, "Intelsat et Intersputnik: Accords Relatifs a L'Exploitation Commerciale des Satellites." *Annuaire Francais de Droit International* 24(1978):890-919.

Courteix, Simone, "Questions d'actualite' en matiere de droit de l'espace." *Annuaire Francais de Droit International* 24(1978):890-919.

Diedericks-Verschoor, I.H. Ph., and W. Paul Gormley, "The Future Legal Status of Non-governmental Entities in Outer Space: Private Individuals and Companies as Subjects and Beneficiaries of International Space Law." *Journal of Space Law* 5:1(1977):125-155.

DeSaussure, Hamilton, "Remote Sensing by Satellites: What Future for an International Regime?" *American Journal of International Law* 71(1977): 707-716.

Doyle, Stephen E. "INMARSAT: The International Maritime Satellite Organization- Origins and Structure," Journal of Space Law 5:1(1977):45-64.

Dula, Arthur, "Free Enterprise and the Proposed Moon Treaty." *Houston Journal of International Law* 2:3(1979):3-53.

Dula, Arthur, "Regulation of Private Commercial Space Activities." *24th Proceedings of the Colloquium on Space law*, International Institute of Space Law, (1981):1-32.

Freibaum, J., and J.E. Miller, "NASA Spectrum and Orbit Utilization Studies for Space Applications." *American Institute for Aernautics and Astronautics*, Paper No. 74-434, 5th Communications Satellite Systems Conference, Los Angeles, 1974.

Galloway, Jonathan F., "The Current Status of the Controversy over the Geostationary Orbit." *21st Proceedings of the Colloquium on Space Law* International Institute of Space Law, (1978):22-27.

Galloway, Jonathan F., "Worldwide Corporations and International Integration: The Case of INTELSAT," *International Organization* 24(1970):503-519.

Gagne, Roland-Yves, "Problemes Juridiques poses par la saturation du Spectre des Frequence et l'encombrement de l'Orbite des Satellites Geostationaires en Matiere de Telecommunications Spatiales." *Revue de Droit de l'Universite de Sherbrooke* 13(Winter 1982):227-254.

Gehrig, James J., "Geostationary Orbit — Technology andLaw." *19th Proceedings of the Colloquium on the Law of Outer Space* 19(1976):267-77.

Goedhuis, D., "Influence of the Conquest of Outer Space on National Sovereignty: Some Observations." *Journal of Space Law* 6:1(1978):37-46.

Goedhuis, D., "The Changing Legal Regime of Air and Outer Space." *International and Comparative Law Quarterly* 27:3(July 1978):576-95.

Gorbiel, Andrezej, "Un Nouveau Probleme du Droit Cosmique International." *Revue Roumanine d'Etudes Internationales* 44(1979):253-262.

Gorove, Stephen, "Freedom of Exploration and Use In the Outer Space Treaty: A Textual Analysis and Interpretation." *Denver Journal of International Law and Policy* 1:1(1971):93-107.

Gorove, Stephen, "The Geostationary Orbit: Issues of Law and Policy." *American Journal of International Law* 73(1979):444-461.

Haanappel, Peter P.C., "Article II of the Outer Space Treaty and the Status of the Geostationary Orbit." *21st Proceedings of the Colloquium on Space Law*, International Institute of Space Law (1979):28-30.

Hallgarten, Katherine D., "The Influence of Communications Satellites on National Communications Laws and Regional Arrangements in the Americas." *Journal of Space Law* 2:1(1974):107-124.

Hill, Arthur, "Domsats Battle for the Arc." *Satellite Communications*, (August 1980), pp. 12-15.

Howkins, John, "The Management of the Spectrum." *Intermedia* 7:5(September 1979):10-22.

Jacobson, Andrew, "Report: Washington, D.C." *Telecommunications*, October 1983, pp. 25-26.

Jakha, Ram S., "The Legal Status of the Geostationary Orbit." *Annals of Air and Space Law* 8(1982):333-352.

Kaltenecker, Hans, "The New European Space Agency." *Journal of Space Law* 5:1(1977):37-43.

Kopal, Vladimir, "The Question of Defining Outer Space." *Journal of Space Law* 8:2(1980):154-173.

Kosuge, Toshio, "National Appropriation of Geostationary Satellite Orbit." *21st Proceedings of the Colloquium on Space Law*, International Institute of Space Law (1979):31-33.

Lay, Fernando, "Space Law: A New Proposal." *Journal of Space Law* 8:1(1980):41-58.

Lesko, Nancy M., "Legal Implications of Direct Satellite Broadcasting—The U.N. Working Group," *Georgia Journal of International and Comparative Law* 6(1976):564-579.

Lipman, Andrew D., "The Rise and Fall of Nineteenth Century Satellite Regulation." *Satellite Communications*, (February 1984), pp. 48-51.

Marsten, Richard B., "Satellites and Space Communication." *Telecommunications Journal* 45(1978):305-314.

Martin, Edward J., and Robert D. Bourne, "Inmarsat: A New Venture in International Cooperation." *Satellite Communications*, (May 1980), pp. 22-23.

McCaskey, Scott D., "Satellite Communications — Stepping Stone for Developing Nations." *Satellite Communications*, (July 1980), pp 22-25.

McKnight, Thomas, and Christopher J. Vizas, "The U.S. Role in International Satellite Communications: The Need to Adapt to Change." *Satellite Communications*, (February 1984), pp. 40-44.

McWhinney, Edward, "The Antimony of Policy and Function in the Internationalization of International Telecommunications Broadcasting." *Columbia Journal of Transnational Law* 13:3(1974).

Menter, Martin, "Commercial Participation in Space Activities." *Journal of Space law* 9:1(Fall 1981):56.

Miles, E., "Transnationalism in Space: Inner and Outer." *International Organization* 25(1971):602-25.

Mizrack, Richard, "The INTELSAT Definitive Arrangements." *Journal of Space Law* 1:2(1973):129-138.

Moore, Amanda Lee, "Direct Broadcast Satellites By Treaty or Regulation: The Committee on Peaceful Uses of Outer Space *v.* the ITU." *19th Proceedings of the Colloquium on Space Law*, International Institute of Space Law (1976):341-349.

Myers, David, "'Common Interest' and 'Non-appropriation' in Outer Space: Political Interpretation of Legal Principles." *International Relations* 6:3(May 1979):529-539.

Pankonin, Vernon, "Protecting Radio Windows for Astronomy." *Sky and Telescope*, (April 1981), pp. 308-310.

Pelton, Joseph, "The Space-Platform Girdle." *Astronautics and Aeronautics*, (May 1981), pp. 42-44.

Perek, L., "Physics, Uses and Regulation of the Geostationary Orbit, or *Ex Facto Sequitur Lex.*" *20th Proceedings of the Colloquium on Space Law*, International Institute of Space Law (1977):400-421.

Rankin, C., "Utilization of the Geostationary Orbit — A Need for Orbital Allocation." *Columbia Journal of Transnational Law* 13:(1976):98-113.

Reifarth, Juergen, "Satellitenfernsehan und Deutsches Rundfunksystem: Vortragsveranstaltung des Instituts fuer Rundfunkrecht an der Universitaet zu Koeln." 14-15 May 1982. Manuscript and notes.

Reijnen, Gijsbertha X.M., "Outer Space Law and Private Enterprise in Outer Space: An International Perspective." *Houston Journal of International Law* 2:15(1979):69.

Robinett, Karen H., "Is Occupation an Appropriation? The Status of a Commercial Facility in Geosynchronous Orbit." *Journal of the British Interplanetary Society* 36(1983):78-80.

Robinson, Glen O., "Regulating International Airwaves: The 1979 WARC." *Virginia Journal of International Law* 21:1(Fall 1980):1-54.

Robinson, Glen O., "The U.S. Faces WARC." *Journal of Communication* 29:1(Winter 1979):150-7.

Rothblatt, Martin A., "International Regulation of Digital Communications Satellite Systems." *Federal Communications Law Journal* 32:3(Summer 1980):393-436.

Rothblatt, Martin A., "The Impact of International Satellite Communications Law Upon Access to the Geostationary Orbit and the Electromagnetic Spectrum." *Texas International Law Journal* 5:16 (1981):207-244.

Rothblatt, Martin A., "International Regulation of Digital Communications Satellite Systems," in *Federal Communications Law Journal* 32(3)Summer 1980:393-436.

Rothblatt, Martin A., "ITU Regulation of Satellite Communications." in *Stanford Journal of International Law*, 18(Spring 1982):1-25.

Rothblatt, Martin A., "Rapid Evolution in Satellite Network Facilities — Legal Implications and the 1985 Space WARC." Paper presented at Forum '83, October 29, 1983, Geneva.

Rothblatt, Martin A., "Satellite Communications and Spectrum Allocation." *American Journal of International Law* 76:1(January 1982):56-77.

Rothblatt, Martin A., "A Jurimetric Framework for the Internaional Allocation and Economic Development of the Orbit/Spectrum Resource." *24th Proceedings of the Colloquium on Space Law*, International Institute of Space Law, (1981), pp. 79-86.

Saint Lager, Olivier, "The Third World and Space Law." *24th Proceedings of the Colloquium on Space Law*, International Institute of Space Law (1981):57-61.

Sawitz, Peter H., "Planning Satellite Communications Services and Spectrum-Orbit Utilization." *AIAA 9th Communications Satellite Conference*, San Diego, March 7-11, 1982, pp. 495-503.

Schultz, James B., "Reliable, Survivable Satellites Seen as Key Link in U.S. National Security." *Satellite Communications*, (June 1980), pp. 26-30.

Skolnikoff, Eugene B., "Science and Technology: The Implications for International Institutions." *International Organization* 25(1971):602-625.

Smith, Delbert D., "Challenge to the 'First-come, First-Served' Principle." *Satellite Communications*, April 1980, p. 6.

Smith, Delbert D., and Martin A. Rothblatt, "Geostationary Platforms: Legal Estates in Space." *Journal of Space Law* (Spring 1982):31-39.

Smith, Delbert D., "International Utilization and Management of Space Systems." *Houston Journal of International Law* (1979):113-129.

Smith, G.K., and G. Benneta, "Geostationary Orbit Capacity in Relation of Services Expansion and Technology Development." *AIAA 9th Communications Satellite Conference*, San Diego, March 7-11, 1982, pp. 495-503.

Sondaal, H.H.M., "The Current Situation in the Field of Maritime Communication Satellites: "INMARSAT." *Journal of Space Law* 8:1(1980):9-39.

Stern, Martin, "Communications Satellites and the Geostationary Orbit: Reconciling Equitable Access with Efficient Use." *Law and Policy in International Business* 14(September 1982):859-883.

Stowe, R.F., "Implications of the 1979 WARC for 12 GHz Satellite Services in Region 2." *23rd Proceedings of the Colloquium on Space Law*, International Institute of Space Law (1980):93-95.

Tong, David, "An Evaluation of Large Space Platforms." *Canadian Aeronautics and Space Journal* 26:4(Fourth Quarter 1980):279-280.

Verschoor-Diederiks, H., and W. Paul Gormley, "The Future Legal Status of Nongovernmental Entities in Outer Space: Private Individuals and Companies as Subjects and Beneficiaries of International Space Law." *Journal of Space Law* 5:1(1977):125-155.

Vilkin, Richard, "Space Law." *California Lawyer*, (February 1982), pp. 28-35.

Williams, Maureen, "The Geostationary Orbit: A Limited Natural Resource from Outer Space." *International Relations* 6:4(November 1979):662-671.

3. ARTICLES ON TELECOMMUNICATIONS

Astrain, Santiago, "Early Bird to INTELSAT IV-A, A Decade of Growth." *Telecommunication Journal* 42:11(1975):45-49.

Astrain, Santiago, "A New Decade for INTELSAT and the Pacific." *Satellite Communications* (April 1980), pp. 16-17.

Atwood, Gary Dean, "WARCS, WAVES, and World Orders — The Politics of the ITU in the Search for New World Information Order." *Journal of International Affairs.* (Fall-Winter 1981-82):267-272.

"Agency Backs Private International Bids." *Aviation Week and Space Technology*, January 9, 1984, p. 63.

Berrada, Abderrazak, "The Effect of the 1979 WARC on the ITU." *Intermedia.* November 1980, pp. 32-33.

Berrada, Abderrazak, "Role of the International Frequency Registration Board." *Telecommunications*, June 1981, pp. 53-59.

Boucher, Eric, "Telecom: la dereglementation americaine ebranle aussi le monopole des P.T.T. en Europe." *Le Monde*, November 8, 1983.

Butler, Richard E., "The ITU's Role in World Telecommunications Development and Information Transfer." *Telephony*, August 22, 1983, pp. 84-86.

Castledine, Susan, "Communication Channels Nearing Saturation Point." *Aviation Week and Space Technology.* March 9, 1981, pp. 102-104.

Codding, George A., "Agenda: Nairobi, 1982." *Telecommunications*, July 1981, pp. 59-61.

Codding, George A., "Telecommunications Expert Criticizes ITU, INTELSAT as Club Dominated by Big Firms." *Communications International*, April 29, 1983.

Codding, George A., "PTT Daggers Drawn for Orion." *The Economist — Connections*, October 27, 1983, pp. 1-2.

Edelson, Burt, and R.D. Briskman, "The Satellite Communications Outlook." *Space Chronicle*, April 1982, pp. 147-155.

Galloway, Eilene, "The United States and International Space Cooperation." Paper prepared for the Office of Technology Assessment, November 1982.

Gregg, Donna C., "Capitalizing on National Self-Interest: The Management of International Telecommunications Conflict By the International Telecommunication Union." *Law and Contemporary Problems* (Winter 1982):37-52.

Hart, Thomas A., "A Review of WARC-79 and Its Implications for the Development of Satellite Communications Services." *Lawyer of the Americas* 12(Spring 1980):442-460.

Honig, David, "Lessons for the 1999 WARC." *Journal of Communications* 30(Spring 1980):48-58.

Hudson, Heather, "WARC '79: Development Communications Strategies —A Report to USAID." *Academy for Educational Development*, Washington, D.C., 1979.

Jackson, Charles, "The Allocation of the Radio Spectrum." *Scientific American*, February 1980, pp. 34-39.

Kirby, Richard C., "Significant Decisions of the CCIR 1982 Plenary Assembly." Paper presented to IEEE International Conference on Communications. Philadelphia, June 1982.

Leive, David M., "INTELSAT in a Changing Telecommunications Environment." Paper presented at Forum '83, Telecom '83, October 29, 1983, Geneva.

Levin, Harvey, "Orbit and Spectrum Resource Strategies — Third World Demands." *Telecommunications Policy* (June 1981):102-110.

Locksley, Gareth, "The Political Economy of Satellite Business." *Telecommunications Policy*, (September 1983):202-205.

Locksley, Gareth, "Legal Issues Up in Air on Satellites." *Los Angeles Times*. March 8, 1983, part VI, pp. 7-8.

Lowndes, Jay C., "ITU Group Promotes Satellites as Aid to Third World Progress." *Aviation Week and Space Technology*. February 4,, 1985, pp. 65-69.

Lowndes, Jay C., "Optical Fiber Threatens Satellite Role in Voice Links." *Aviation Week and Space Technology*. January 31, 1983, pp. 62-65.

Lowndes, Jay C., "Conference Sets Bands, Orbital Slots." *Aviation Week and Space Technology*. August 8, 1983, p. 49.

McLauchlan, William P. and Richard M. Westerberg, "Allocating Broadcast

Spectrum: Models and Proposals." *Telecommunications Policy*, June 1982, pp. 111-122.

Melody, William H., "Economics of Spectrum Allocation." Paper presented at Pacific Telecommunications Conference, Honolulu, January 1983.

Mizrack, Richard, "The INTELSAT Definitive Arrangements." *Journal of Space Law* 1:2(1973):129-138.

Pelton, Joseph, *et al.*, "INTELSAT: The Global Telecommunications Network". Monograph presented at the Pacific Telecommunications Conference, Honolulu, January 1983.

Pelton, Joseph, *et al.*, "INTELSAT: The Geosynchronous Orbital Arc and the Third World." Paper presented at Pacific Telecommunications Conference, Honolulu, January 1983.

Rutkowski, A.M., "Six Ad-Hoc Two: The Third World Speaks Its Mind." *Satellite Communicaitons*, March 1980, pp. 22-27.

Rutkowski, A.M., "The 1985 Space WARC Advisory Committee." *Telecommunications*, November 1981, pp. 81-82.

Rutkowski, A.M., "The Impact of New Technology on Satellite Radiocommunication." *Telecommunications*, February 1983, Paper submitted to FCC Space WARC Advisory Committee.

Rutkowski, A.M., "The Inter-Satellite Service: Future Shock for Telecommunication Policy Makers." *Telecommunications*, July 1981, pp. 64-65.

Rutkowski, A.M., "The Space WARC." *Telecommunications*, (January 1984), source: "Teleclippings," March 12, 1984, pp. 1-2.

Rutkowski, A.M., "United States Policymaking for the Public International Forums on Communications." *Syracuse Journal of International Law and Commerce* 8:1(1981):98-143.

Segal, Brian, "International Negotiations on Telecommunications." *Intermedia* 8(November 1980):21-31.

Vicas, Alex G., "An Economic Assessment of the CCIR's Five Methods for Assuring Guaranteed Access to the OSR." *Annals of Air and Space Law* 7:(1983):431-446.

Warner, Margaret G., "U.S. Telecommunications Deregulation is Causing Plenty of Static from Europe." *The Wall Street Journal*, November 29, 1982, p. 12.

SELECTED INTERVIEWS

Barberis, Neil, System Engineer, Space Systems Operation, Ford Aerospace and Communications Corporation, Palo Alto, California, February 6, 1984.

Butler, Richard E., Secretary-General of the International Telecommunication Union. Interviews at UNISPACE '82, Vienna, August 1982; ITU Headquarters, Geneva, December 1982: Pacific Telecommunications Conference, Honolulu, January 1983; and Telecom '83, Geneva, October 1983.

Christol, Carl Q., Professor of Political Science and International Law, University of Southern California. Interviews in Los Angeles, on January 25, 1982; at Western Political Science Association Conference (see below), March 28, 1982; and at the Proceedings of the Colloquium on Space Law, Budapest, October 9-15, 1983.

Clarke, Arthur C., Writer and Scientist, originator of the idea for communications satellites in the geostationary orbit. Interview at UNISPACE '82, August 10, 1982, Vienna.

Dal Bello, Richard, Policy Analyst for U.S. Congress, Office of Technology Assessment. Interviews at UNISPACE '82, Vienna, August 1982; OTA Workshop and Advisory Group Meetings, Washington, D.C., December 1982.

Hilewick, Carol Lee, Director of United States World Communications Year Committee. Interview at Pacific Telecommunications Conference, Honolulu, January 1983.

Jansky, Donald M., Assistant Director, National Telecommunications and Information Agency, Washington, D.C. Interview at UNISPACE '82, Vienna, August 1982.

Kirby, Richard, Director of the International Radio Consultative Committee (CCIR), International Telecommunication Union. Interview on May 29, 1982, Geneva.

Lundberg, O., Director-General of INMARSAT, UNISPACE '82, August 10, 1982.

Lusignan, Bruce B., Director, Communications Satellite Planning Center, Stanford University, Palo Alto, California, February 7, 1984.

Macuk, David, Telecommunications Attache, United States Mission to the United Nations, Geneva. Interviews on May 28, 1982 and December 9, 1982.

Meyerhoff, Henry, Counsellor to International Frequency Registration Board, International Telecommunication Union. Interview on May 28, 1982, Geneva.

Nickelson, Richard, Senior Counsellor to International Radio Consultative Committee, International Telecommunication Union. Interviews on May 29, 1982; and UNISPACE '82, Vienna, August 1982.

Sant, M., International Frequency Registration Board, International Telecommunication Union. Interview on January 14, 1985, Pacific Telecommunications Conference, Honolulu.

Smith, Marcia, Policy Analyst for the Congressional Research Service, Library of Congress. Interviews at UNISPACE '82, Vienna, August 1982; and International Astronautical Federation Congress, Budapest, October 1983.

Tycz, Thomas, International Staff, Federal Communications Commission. Interview on December 3, 1982, Washington, D.C.

CONFERENCES ATTENDED

American Institute of Aeronautics and Astronautics (AIAA), 9th Communications Satellite Systems Conference, San Diego, California, March 1982.

Western Political Science Association Conference, Panel on Outer Space Law, San Diego, California, March 28, 1982.

CCIR, Conference Preparatory Meeting, Geneva, June 28, 1982.

United Nations Conference on the Exploration and Peaceful Uses of Outer Space (UNISPACE '82), Vienna, August 9-21, 1982. Non-governmental Organization Conference and Exhibitions held concurrently.

Hughes Aircraft "Big Bay." Tour of satellite construction and test facilities at Hughes Aircraft, El Segundo, California, October 28, 1982.

Office of Technology Assessment, Workshop and Advisory Panel Meeting, UNISPACE '82 Technical Memorandum, November 30 and Decdmber 2, 1982, Washington, D.C.

Pacific Telecommunications Conference, Honolulu, January 15-19, 1983.

AIAA Aerospace Engineering Conference, Panel on Space Law, Long Beach, California, May 11, 1983.

International Student Pugwash Conference on Science and Technology, University of Michigan, Ann Arbor, Michigan, June 15-21, 1983.

International Astronautical Federation Congress, International Institute of Space Law Colloquium on Space Law, Budapest, October 9-15, 1983.

4th Telecommunications Exhibition (Telecom '83), Forum '83 on Legal Questions dealing with International Telecommunications Technologies, Geneva, October 26-November 2, 1983.

International Communication Association Conference, San Francisco, May 24-28, 1984.

International Astronautical Federation Congress, International Institute of Space Law Colloquium on Space Law, Lausanne, October 7-13, 1984.

International Business in Space Conference Washington, D.C., January 9-11, 1985.

Pacific Telecommunications Conference, Honolulu, January 13-17, 1985.